S. Hrg. 106–415

PADUCAH GASEOUS DIFFUSION PLANT

HEARING

BEFORE A

SUBCOMMITTEE OF THE
COMMITTEE ON APPROPRIATIONS
UNITED STATES SENATE

ONE HUNDRED SIXTH CONGRESS

FIRST SESSION

SPECIAL HEARING

Printed for the use of the Committee on Appropriations

Available via the World Wide Web: http://www.access.gpo.gov/congress/senate

U.S. GOVERNMENT PRINTING OFFICE

61–413 cc　　　　　　　　　　WASHINGTON : 2000

COMMITTEE ON APPROPRIATIONS

TED STEVENS, Alaska, *Chairman*

THAD COCHRAN, Mississippi
ARLEN SPECTER, Pennsylvania
PETE V. DOMENICI, New Mexico
CHRISTOPHER S. BOND, Missouri
SLADE GORTON, Washington
MITCH McCONNELL, Kentucky
CONRAD BURNS, Montana
RICHARD C. SHELBY, Alabama
JUDD GREGG, New Hampshire
ROBERT F. BENNETT, Utah
BEN NIGHTHORSE CAMPBELL, Colorado
LARRY CRAIG, Idaho
KAY BAILEY HUTCHISON, Texas
JON KYL, Arizona

ROBERT C. BYRD, West Virginia
DANIEL K. INOUYE, Hawaii
ERNEST F. HOLLINGS, South Carolina
PATRICK J. LEAHY, Vermont
FRANK R. LAUTENBERG, New Jersey
TOM HARKIN, Iowa
BARBARA A. MIKULSKI, Maryland
HARRY REID, Nevada
HERB KOHL, Wisconsin
PATTY MURRAY, Washington
BYRON L. DORGAN, North Dakota
DIANNE FEINSTEIN, California
RICHARD J. DURBIN, Illinois

STEVEN J. CORTESE, *Staff Director*
LISA SUTHERLAND, *Deputy Staff Director*
JAMES H. ENGLISH, *Minority Staff Director*

———

SUBCOMMITTEE ON ENERGY AND WATER DEVELOPMENT

PETE V. DOMENICI, New Mexico *Chairman*

THAD COCHRAN, Mississippi
SLADE GORTON, Washington
MITCH McCONNELL, Kentucky
ROBERT F. BENNETT, Utah
CONRAD BURNS, Montana
LARRY CRAIG, Idaho
TED STEVENS, Alaska *(ex officio)*

HARRY REID, Nevada
ROBERT C. BYRD, West Virginia
ERNEST F. HOLLINGS, South Carolina
PATTY MURRAY, Washington
HERB KOHL, Wisconsin
BYRON DORGAN, North Dakota

Professional Staff
ALEX W. FLINT
W. DAVID GWALTNEY
GREG DAINES *(Minority)*

Administrative Support
LASHAWNDA LEFTWICH

CONTENTS

PADUCAH GASEOUS DIFFUSION PLANT

TUESDAY, OCTOBER 26, 1999

U.S. SENATE,
SUBCOMMITTEE ON ENERGY AND WATER DEVELOPMENT,
COMMITTEE ON APPROPRIATIONS,
Washington, DC.

The subcommittee met at 9:36 a.m., in room SD–124, Dirksen Senate Office Building, Hon. Pete V. Domenici (chairman) presiding.

Present: Senators Domenici, McConnell, and Craig.

PADUCAH GASEOUS DIFFUSION PLANT WORKFORCE SAFETY AND EXPOSURE ISSUES

OPENING STATEMENT OF PETE V. DOMENICI

Senator DOMENICI. The hearing will please come to order.

Governor PATTON. Am I supposed to be with the next panel, Senator?

Senator MCCONNELL. Governor, why do you not just stay. It is fine.

Senator DOMENICI. Good morning, everyone, and welcome to those who are going to testify, those who are here to observe, and welcome, members of the press, to this subcommittee hearing.

This is the Energy and Water Development subcommittee, kind of a misnomer in the sense it is charged with funding the water projects of this country, but on the other side of the ledger it is charged with funding all of the nuclear activities that pertain to the defense of our Nation along with the Department of Energy's non-nuclear research projects. It has fallen to this subcommittee over time to try to handle the cleanup in the nuclear waste activities of our Nation since the origin of the atomic bomb.

There is no doubt that this morning the committee meets to consider environment, safety, and health issues associated with the operation of the Paducah uranium enrichment plant while it was operated by the Department of Energy. Senator McConnell has shown real leadership in putting together this hearing. The witnesses are experts in their fields and I think the subcommittee is going to be very well informed after today's hearings.

I want to assure both Senators, Senator Bunning and Senator McConnell, and the others here that the committee takes its obligation of clean up very seriously. We are massively oversubscribed in that regard. There is more to do than can possibly be done in the foreseeable future. However, the subcommittee has shown real enthusiasm for this—the establishment of priorities that will ensure that the Federal Government meets its obligations.

As Senator McConnell knows, I will not be able to stay for the entire hearing this morning, but I am delighted to open it and will be here as long as I can.

With that, I am now going to welcome Senator Bunning, Governor Patton, Congressman Whitfield, and I am going to yield to Senator McConnell, who will preside over the hearing.

Senator McConnell.

STATEMENT OF MITCH MC CONNELL

Senator McCONNELL. Well, thank you, Mr. Chairman. I appreciate very much your making this subcommittee hearing today possible.

Today, the Energy and Water Subcommittee will investigate the reports of the Department of Energy's failure to properly protect the safety of the work force and the environment at the gaseous diffusion plant in Paducah. It is the goal of the subcommittee to gain a clear understanding of what has occurred, in many cases what has not occurred, and what must be done to properly accelerate cleanup, protect worker safety, and identify the health problems related to exposure to plutonium.

I would also like, as I indicated earlier, to thank again Senator Domenici for giving me an opportunity to chair the hearing today.

WORKPLACE HEALTH AND SAFETY

On August 8, the Washington Post ran a series of stories based on very serious allegations that the Department of Energy used recycled nuclear fuel that was laced with plutonium and other radioactive material without informing the work force that handled this highly toxic material. As a result, an estimated 15,000 former workers and 5,500 current workers at the three gaseous diffusion plants have been put in harm's way.

In the intervening 3 months, much more has come to light about what happened at the Paducah plant. We also have uncovered several documents dating from the 1950's, when production first began at the plant, that identify problems in the area of worker protection and DOE's failure to disclose these findings to the workers.

Whether it was a 1952 memo acknowledging that plutonium is a greater hazard than uranium, a March 1960 document identifying the presence of neptunium, which is highly radioactive, and the Department's efforts to conceal this information from workers, a 1980 report by the Comptroller General finding that Oak Ridge operations had failed to effectively implement its health and safety program, or a 1990 tiger team report which identified many of the same regulatory errors that recently have been identified again 10 years later in DOE's phase one investigation, what is clear from all these documents is the half century of dangerous activity and deception in Paducah.

From reading this record, it is clear that DOE at worst lied about the presence of harmful contaminants and at best covered up this information, both of which are unacceptable. It is abundantly clear from these reports that the Department is unable to adequately perform its job as site operator and independent regulator. It is like putting a mouse in charge of the cheese.

Changes must be made. They must be made now. One place to start is that we move forward and monitor workers for illness possibly connected with their work at the enrichment facilities.

SITE CLEANUP

During the Paducah field hearing on September 20, 1990, the DOE site manager testified that the Department spends 89 percent of its annual cleanup budget in meeting the existing environmental standards, leaving just 9 percent going toward future cleanup. Let me repeat: For every one dollar spent in Paducah, about a dime goes to clean up the mess, 90 percent is spent on paperwork and regulation not directly related to solving the problem.

For the last 10 years the Department has spent $400 million and done little to remove the worst contamination. Today our goal is clear: We need to insist that DOE spend future appropriations on the elimination of contamination, not on the creation of paperwork.

CLEANUP AND REGULATORY IMPROVEMENTS NEEDED

While I support most of the modest reforms identified in the phase one investigation, I believe the Department must consider a number of more substantive and bold reforms. These measures will shake up the bureaucratic chain of command and may help put an end to the constant tug of war within DOE that has hindered the flow of funding for Paducah and hampered the progress of the cleanup.

Careful consideration must be given to establishing an independent regulator at the Paducah and Portsmouth sites to ensure that worker and environmental protection always remains a priority. Foremost among the Department's priorities should be the expansion of the worker health testing program, including providing important lung screening that can assist in identifying early stages of lung cancer.

The administration also must do more to tap the available surplus in the uranium decommissioning and decontamination fund to ensure that the revenue collected for cleanup actually goes to cleanup. After reviewing the phase one investigation, it is clear to me that very little has actually been cleaned up, despite the Department's having spent $400 million of the taxpayers' money.

Further, it is clear that DOE has failed to prioritize cleanups on risk. This has resulted in budget requests that fail to address areas like Drum Mountain, which is depicted in the chart that we have back here. This is Drum Mountain. That is a picture of it today.

STATUS OF URANIUM D&D FUND

It is also important to note that the current balance in the uranium cleanup trust fund, which provides funds for the cleanup of the three gaseous diffusion plants, currently maintains a $1.6 billion surplus. Based on current spending, current spending projections, this account is estimated to grow to $6 billion by 2007. Considering the massive surplus in the trust fund, the administration must make the cleanup in Paducah a priority within its overall budget, as well as set a new corrective action plan for Paducah and Portsmouth that will accelerate cleanup at the sites.

CLEANUP PRIORITIES

Another important reform DOE must consider is allowing Paducah and Portsmouth to set their own cleanup goals and objectives, instead of accepting the priorities set several hundred miles away in Oak Ridge, TN. Based on the information contained in the phase one investigation, it is apparent that Oak Ridge has been negligent in its oversight and has not been responsive to cleanup needs or to protecting worker health and safety.

I think a good example of the Department's inattention to risk cleanup is best illustrated by an October 30, 1996, project manager's meeting notes that I will have included in the record. The notes document the debate between DOE and Tennessee State regulators on whether or not to spend $250 million to clean up three buildings in Oak Ridge, even though they are not a risk-based priority. This year the Department will spend $62 million to clean up these buildings, which is $25 million more than DOE had budgeted for the entire Paducah effort.

It is examples like these that demonstrate DOE's poor judgment when it comes to assessing cleanup priorities. By empowering Paducah and Portsmouth to make their own cleanup decisions I believe much more can be accomplished in a shorter amount of time.

We look forward to hearing from the witnesses today to determine whether these moderate reforms proposed by DOE and the more substantive reforms I have raised for consideration will achieve the goals that I have laid out, to make worker testing a priority and to accelerate cleanup.

NATURE OF PADUCAH CONTAMINATION

But before we get started, I just want to explain what we have behind us. The chart 1 here is of Drum Mountain. This is a site which covers five acres, contains five concentrations of radioactive contamination, and continues to contribute to soil and groundwater contamination. Every time it rains, it washes more and more radioactive contamination into the groundwater, which further complicates DOE's cleanup efforts.

In the second picture, a little further over, is a radioactive black ooze that was haphazardly uncovered by DOE's staff when they were investigating offsite contamination. The material was found nearby the so-called sanitary landfill. If a picture says a thousand words, this one has to say a lot about DOE's level of protection. You will notice here that this stand is not even upright, not that that would provide much protection anyway or much notice to people to stay away from it. But the stand is even knocked over. And this is the black ooze.

As we have indicated, these are the drums that the Governor, Senator Bunning, and Congressman Whitfield and I are all too familiar with because we have seen them.

Senator DOMENICI. Thank you. I will stay for just a while longer.

I want to also recognize that we have Senator Bunning and Congressman Whitfield here, Members of Congress, and in their instances, in their respective bodies and committee work, they have indicated a serious and abiding interest in trying to resolve this. I understand Senator Bunning as a member of the Energy and Natural Resources committee has already conducted a hearing in the State, and we welcome you here today.

Governor, with your permission we will start with Senator Bunning and then we will go to Congressman Whitfield, and then we will welcome you for your remarks.

Senator Bunning.

STATEMENT OF HON. JIM BUNNING, U.S. SENATOR FROM KENTUCKY

Senator BUNNING. Thank you, Senator Domenici and Senator McConnell. I appreciate you holding this hearing today, focusing some very much needed attention on the problems surrounding the gaseous diffusion plant in Paducah, KY, and hopefully starting us on the path to some solutions to the problems.

Last month on September 20, at my request the Senate Committee on Energy and Natural Resources held a field hearing in Paducah. Our hearing focused primarily on gaining some input from the people who have been directly involved at the gaseous diffusion plant, the workers. I would like to give you a brief synopsis of what we heard from those workers during that hearing. I think their testimony clearly demonstrates how serious this problem is.

HUMAN IMPACTS AND EXPOSURES

For example, we heard from Mr. Eugene Stallings, who used to work in what they called the C–410 feeding plant building, where used reactor fuel rods were ground up and mixed with fluorine to start the uranium enrichment process. Or, as Mr. Stallings put it, "this is where they made the hot stuff."

One of Mr. Stallings' jobs was to make sure that the pipes carrying this dangerous material did not get plugged. Periodically, however, according to Mr. Stallings, some of the lines would plug and you would have to use a rod to punch out the frozen material, sometimes for as long as 8 hours working in these pipes with the hot stuff.

It was later determined that the hot stuff Mr. Stallings worked with, without benefit of a respirator, was 700 times more radioactive than a person should have been exposed to.

Mr. Stallings was followed by D.R. Johnson, a former welder at the plant. Mr. Johnson was required to work in an environment that was routinely filled with thick smoke and dust, that for the most part included asbestos and PCB's. However, he was never provided a respirator or protective clothing. In fact, Mr. Johnson told us how he was taken off the line because at one point he tested hot.

Nobody explained to him what being "hot" actually meant. He asked, but he was told there was always someone else that could do his job and that if he did not like it he could hit the road.

We also heard from Mr. Michael Roberts, who worked in the 410 feeder building. He told the committee that at the time the proper equipment for changing filters was considered to be a jacket taped at the cuffs, bath towels stuffed around the collar and draped over your head, and World War II type gas mask. "It was the only thing we could do to keep the powder off our skin."

Finally, Mr. Philip Foley, a 24-year veteran of the plant, told us that when he first started working at the plant they would dispose of contaminated waste by pouring it into barrels and tossing them into ponds. At the time those barrels would burst into flames, creating huge plumes of contaminated smoke. But they were told not to worry and to just throw dirt on the fire and cover it up.

As worker after worker testified, it became clear that this was not just a few disgruntled employees blowing the whistle on a couple of bad managers. It became clear that the horror stories of inadequate safety procedures or equipment and improper, haphazard disposal of hazardous wastes were not isolated incidents, but that they were prevailing standard operating procedure at this plant for many years, endangering the health of the workers at the plant and jeopardizing the health of the neighboring community.

The frightening thing is that we do not know even now the extent of the problem. We do not know where the waste was buried. We do not know where all the ponds are that had barrels dumped in them.

DOE 1990 INVESTIGATION

We do know this: In 1990 the Department of Energy sent a so-called tiger team to investigate reports of environmental problems at the plant. What they found was an area devastated by years of unsafe dumping, with possible radioactive wastes seeping into the drinking water supply and workers inadequately trained and protected from radioactive waste.

That tiger team report must have been rather toothless, because now, nearly 10 years later, the Department of Energy phase one investigation reports that there is still radioactive waste seeping into the water supply, that the workers are still not being provided with adequate amounts of training and equipment, and that we still do not know where all the waste might be buried.

Ten years have gone by, $400 million has been spent, and nothing has changed. Not one contaminated drum has been removed. Not one ounce of spent uranium has been converted. And the plume of contaminated waste that includes PCB's continues to flow toward the Ohio River.

Mr. Chairman, something needs to change. We cannot wait another day to do it. The workers in this plant have been betrayed. The community which supported this facility has been betrayed. They trusted the U.S. Government. It is time to provide the resources to clean up this mess, to provide health care benefits to those who need it, and to correct the environmental damage that has been done.

Some experts estimate that it will cost nearly $1 billion to clean up the Paducah site. I know this is a large sum of money, but after touring the plant and seeing the mountains of contaminated drums, the acres of canisters filled with dangerous spent uranium, and following the plume of waste that is spreading in the area's drinking water and the Ohio River, it is time to deliver.

It is a lot to ask for, but hopefully after today's hearing you too will recognize what we are up against and provide the necessary resources to begin the cleanup process in Paducah. Mr. Chairman,

I urge you and the rest of your colleagues on the committee here today to do just that.

I thank you for the time that you have given me.

Senator DOMENICI. I am now going to yield the chair to the distinguished Senator from Kentucky. Thank you very much.

Senator MCCONNELL [presiding]. Thank you, Mr. Chairman.

Thank you, Senator Bunning, not only for having the first hearing in Paducah on the 20th, but being here today.

I also want to at this point call on Congressman Whitfield, who is a member of the House Commerce Committee and the Subcommittee on Oversight and investigations, which held a hearing on September 22 over on the House side. Ed and Jim have both been deeply involved in this and I want to give Congressman Whitfield an opportunity now to bring us up to date on his thoughts on the subject.

STATEMENT OF HON. ED WHITFIELD, U.S. REPRESENTATIVE FROM KENTUCKY

Mr. WHITFIELD. Well, Senator McConnell, thank you very much. I am delighted that this hearing is being held. I would just say to you that it has been quite helpful working with your staff on this issue, with Senator Bunning's staff, as well as people from the State of Kentucky, Governor.

I will simply comment on the focus of the September 22 hearing held by my Subcommittee on Oversight and Investigations and share my general observations on what has happened in the past and what should happen in the future. The House hearing had three panels of witnesses. We heard from the relaters in the lawsuit filed against one of the former plant contractors under the False Claims Act. We heard also from representatives of the former and current contractors, as well as employees at the plant. And we heard from Federal and State regulators responsible for overseeing work at the production facility and in the area of environmental management and cleanup.

After a full day of testimony and a series of meetings, I have reached personally the following conclusions. The Department of Energy has been deficient in overseeing the work of prime contractors at the site. I will say that the Department of Energy at this time seems to be recognizing that, acknowledging that, and moving forward and trying to solve some of these problems.

The prime contractors have failed to properly protect the health and safety of the workers and, although the prime contractors have changed periodically and there is a perception that change is taking place, frequently the management team of those contractors has not changed, and as a result some of the same mistakes continue to be made.

DOE CLEANUP EFFORT AT PADUCAH

As was stated in your opening statement, the vast majority of environmental cleanup funds are spent simply to comply with existing regulations, thereby resulting in very little actual cleanup. As a matter of fact, nationwide last year $5.8 billion was appropriated for cleanup of the DOE sites around the country and over $3.5 billion of that was spent in compliance alone. So not only is this occurring in Paducah, but nationwide very little money is going for actual cleanup.

Cleanup efforts at Paducah and Portsmouth should not be managed, in my view, by decisionmakers at Oak Ridge, TN. Last year $240 million was appropriated for cleanup, and of that Paducah received $37 million. The Paducah and Portsmouth plants in my view are not being treated in an equitable manner in the distribution of cleanup funds.

Generally speaking, I have also concluded there have been serious operational deficiencies that continue to be within DOE's area of responsibility. Workers have not been properly notified of potential threats to their health and safety. They have not been properly clothed, equipped, or monitored when exposed to dangerous hazardous materials. Records of exposure to radioactive and toxic substances are incomplete and inaccurate. Environmental hazards have not been properly characterized even today. Remediation of existing contamination is too slow and costly and funding shortfalls also have slowed our progress.

COMPENSATION PROGRAM

As I stated from the outset of this controversy, our number one priority, my number one priority, is the health and safety of the workers at the plant and the citizens in the surrounding communities. I believe it is imperative that we adopt legislation to establish a Federal compensation program for employees who have suffered illness as a direct result of the exposure to these radioactive materials.

Secretary Richardson recently announced that they were expanding the program to take care of workers who had been exposed to beryllium. We should do the same thing for those exposed to radiation.

In addition, I would say that the medical monitoring program which is now in existence should be extended to current workers as well as past workers, and that the most recent technology, such as CT scanners, should be used to help determine if some of these workers have cancer or other illnesses.

I would also say that later this morning you are going to hear from the president of the local PACE union at Paducah, David Fuller, his associate Jimmy Keyes. They have been quite valuable to all of us in this process. It has been one of those issues where everyone could become quite emotional about it and overreact, and I know that David and his associates have been under pressure, and they have reacted in a calm manner, trying to come forward with constructive solutions, and I am delighted that they will be testifying today.

Thank you very much, Senator, for giving me the opportunity to testify this morning. I am delighted that we are focusing on this issue. It is a very serious issue and I know that we can come up with some solutions.

Senator MCCONNELL. Thank you, Congressman Whitfield, for your aggressive action across the board on this important subject.

I just might say to our witnesses, Senator Bunning and I have a vote at 10:00 o'clock and I think the best way to do this would be to excuse Senator Bunning and Congressman Whitfield and,

Governor, Jim and I will go vote, and then I will be back and then I will be back and take your testimony. I think that way we will not have to interrupt what you have to say.

So the hearing is recessed while I go vote.

[A brief recess was taken.]

Senator MCCONNELL. The hearing will come to order once again.

I want to apologize to the Governor for having to run vote, but that happens from time to time. We are very happy to have you here, Governor. I know you have been very active having been down to the plant several times. We are anxious today to get your view from the perspective of the Commonwealth of Kentucky on the cleanup issue at the plant. I want to thank you very much for the leadership you have shown and welcome you here today, and look forward to hearing from you.

STATEMENT OF HON. PAUL PATTON, GOVERNOR, COMMONWEALTH OF KENTUCKY

Governor PATTON. Thank you, Mr. Chairman for your remarks, and thank you for the opportunity to discuss with you the problems at the Paducah gaseous diffusion plant and to ask that this subcommittee work with the Commonwealth of Kentucky to ensure that the Federal Government honors its moral obligation and contractual commitment to clean up the contamination in this area by the year 2010. I have submitted a more detailed statement for the record and I will try to summarize it for you now.

Mr. Chairman, it is time for the Federal Government to do right by the city of Paducah, a city that has been loyal to this plant for 47 years. When the allegations contained in the Federal whistleblower lawsuits first began to draw national attention in August, I asked my staff and cabinet to report to me whether we were currently doing all we could to protect the workers at the plant and the health of the general public and the environment in the area.

Like yourself, Senator Bunning and Congressman Whitfield, I have personally toured the site in August and spent some time with the workers to see if they felt comfortable with the safety procedures that are in place at the plant currently. At that time these workers told me they felt safe, but were concerned about what may have happened in the past.

I have also concluded that at the present time the State is doing all it can to contain threats to the general public's health and safety and is doing all we can to monitor compliance with accepted environmental practices.

DOE CLEANUP PRIORITY AND SCHEDULE

But the problem gets worse every day it is not addressed. Despite the fact that I found no current danger to public health in the region, my administration's efforts have led me to one obvious and inescapable conclusion: This site is one of the most environmentally contaminated in the South, and the Federal Government is not devoting the necessary funds to meet its obligation to clean it up.

In 1994 this site was placed on the national priorities list, and after that designation our natural resources and environmental protection cabinet entered into an agreement with the U.S. Department of Energy and the Environmental Protection Agency whereby the Department of Energy agreed to fund and complete the cleanup of the site by the year 2010. We wanted and felt it would be reasonable to have this work done by 2007, but in an effort to be cooperative we accepted the later date.

But we now discover that, based on the current rate of progress, it will not be cleaned up in our lifetime. I will leave the details of the contamination at the site to later panels of regulators, but I can assure you that we have now determined that the situation is more serious than we first thought. This subcommittee needs to know that this is a site with, as you say, acres of radioactively contaminated waste materials and scrap and metal piles that you have illustrated, open ditches contaminated with elements like plutonium, a radioactive underground water plume moving toward the Ohio River at an alarming rate, and 37,000 cylinders of depleted uranium stored outdoors, exposed to the elements, and inadequately protected from deterioration.

As I have learned more about the environmental hazards at the site, I have become most alarmed, not by the extent of the contamination, which is very alarming, but by the fact that the Department of Energy currently does not have, nor does it plan to request in the near future, sufficient funds to address these serious environmental dangers. Mr. Chairman, the people of Paducah and the lower Mississippi River Valley deserve better than that.

Our best estimate is that it will require at least $200 million a year for the next 10 years to address this issue. The Department of Energy has planned budget requests totaling only $630 million through fiscal year 2010, far short of the $2 billion that we estimate this project will cost.

Even more disturbing, these inadequate projected requests anticipate huge funding increases beginning in 2007. Their projections for the next 7 years average less than $50 million a year.

Environmental management funding at Paducah has been about $38 million a year over the last several years, and of this $38 million only about $11 million per year has been actually going to environmental remediation at the site. They are not planning to do much more during the next 7 years than they are already doing, and it will be impossible physically and financially to cram this much remediation into the last 3 years of the agreement.

Lacking detailed facts, our estimates are just that. But do not take our word for it. As you illustrated with this phase one independent study, it illustrates that they have admitted that the current cleanup schedule is totally unrealistic based on the current funding levels and, two, the estimated cleanup costs are based on faulty assumptions, such as unproven technologies and leaving hazardous materials on site, which is not acceptable to us.

It is time for the Department of Energy to reassess the costs of this cleanup and to be forthcoming about the true projected costs. Mr. Chairman, the Congress has already made provisions to fund this cleanup. The Department of Energy environmental management activities at the site are funded, as you mentioned, from the Uranium Enrichment Decontamination and Decommissioning Fund. The total appropriations from this fund for fiscal year 1999 was about $220 million, of which Paducah received about $36 million.

We believe that this is not a fair or rational division of that $220 million, and it disturbs me that the responsible officials believe that Oak Ridge should get over 60 percent of this money and Paducah less than 20 percent.

But even more disturbing is the fact that the D&D fund takes in almost $610 million a year, as you noted, and only $220 million is appropriated for its intended use. The D&D fund has a positive balance, again as you mentioned, of $1.5 billion. Mr. Chairman, it is time for the Federal Government to accept responsibility for the problem and to begin to eliminate it.

As I have discussed with you previously, I am asking the Congress, the Department of Energy, OMB, and the White House to immediately appropriate an additional $100 million to the cleanup at Paducah so we can adequately document the problem and begin the cleanup in a serious way. Only a figure of this magnitude can get us moving toward completing the cleanup by 2010.

I have already informed the administration if they are going to be an environmental administration in a regulatory fashion, passing the costs along to the customers of private companies, then they must also be an environmental administration in a matter of Federal financial responsibility.

I have in the strongest terms urged the administration to ask for enough funds to do this job. If they do, I ask the Congress to approve it. If they do not, I ask the Congress to ensure that our government keeps its commitments to the people of the region affected by this problem.

I call upon the Congress to find a way to work with the Department of Energy to fully fund the D&D program for its intended purposes and to make certain that funds are available to complete the cleanup at Paducah by 2010.

I stand ready to work with the White House, the Congressional delegation, and the political leadership of both parties in this effort. But I am determined to get the process accelerated and to see to it that the agreement reached last year is implemented. As a result of signing the Federal Facilities Agreement, the Commonwealth now has several legal means at its disposal to ensure that the cleanup proceeds in a timely manner. If current funding levels are maintained, the Commonwealth believes that DOE will be in default of that agreement as early as fiscal year 2001.

Let me assure you, the people of Paducah and Kentucky that I will continue my efforts on this issue and that our administration will use every political or legal means at our disposal to make certain that the obligations of the Federal Government are met. We can in good conscience do no less.

Thank you, Mr. Chairman, for your attention. Let me add to the record our estimation of—and I do not believe this was in my testimony—our comparison of the DOE's estimate and our estimate, differs by about $1.2 billion. I would like to add that to the record.

Senator MCCONNELL. I appreciate that, Governor. We will make that a part of the record.

Governor PATTON. And I would like to note that this is a three-State effort. I have a letter which is being mailed to all of the members of the subcommittee from the Governors of Kentucky, Tennessee, and Ohio. That should have been mailed or will be mailed to each member of the committee.

PREPARED STATEMENT

With that, Mr. Chairman, thanks for your attention and I would be glad to answer any questions.
[The statement follows:]

PREPARED STATEMENT OF GOV. PAUL PATTON

Mr. Chairman, members of the subcommittee, I appreciate the opportunity to appear before you today to highlight the ongoing environmental concerns in and around the Paducah Gaseous Diffusion Plant and to ask that this committee work with the political leadership of the Commonwealth of Kentucky to ensure that the federal government honors its moral obligation and contractual commitment to cleanup the contamination in that area by the year 2010.

The Paducah Gaseous Diffusion Plant was opened in 1952 and has been in operation since that time. It initially processed nuclear materials for the military, but in the mid 1960s, its mission shifted to the commercial focus of enriching uranium for use in nuclear reactors. The 800 acre plant is located on approximately 3,400 acres of federal land about three miles south of the Ohio River and twelve miles west of Paducah, and has been the largest employer in the area since the 1950s.

From an economic standpoint, the plant has been good for Paducah. It has provided many, good-paying jobs to the region. It has been a stable force in the local economy. And in turn, McCracken County and the City of Paducah have been good to the federal government. They have accepted the uranium enrichment complex in their region and have valued it as an employer. Paducah has proven itself to be a city that is tolerant of this activity, and its population has become educated on the uranium enrichment process and has learned to separate legitimate concerns from exaggerated fears. Paducah has stood well by the federal government in this effort.

Now it is time for the federal government to do right by Paducah.

When the allegations contained in the federal whistleblower lawsuits first began to draw national attention in August, I asked my personal staff, led by Jack Conway, to work with our cabinets in Kentucky State Government and report to me on whether the Commonwealth had been negligent in the past or whether we were currently doing all we could to protect public health and the environment in the area. Like Senators McConnell and Bunning, and Congressman Whitfield, I personally toured the site in August and spent time with some of the workers to see if they felt comfortable with the safety procedures that are in place at the plant. While visiting with the workers, I heard that, by and large, they felt well trained for the materials they handle, and had general confidence in the safety procedures currently in place at the facility. Some expressed concerns about what they had heard of past practices, but they felt generally positive about current safety.

After consulting with the Kentucky state agencies responsible for monitoring the environment and public health, I concluded that Kentucky is doing all we can presently do to contain threats to public health and all we can presently do to monitor compliance with accepted environmental practices—although we have been prevented by the federal government from monitoring these activities like we would have had this operation been conducted by a non-governmental entity. In August 1999, our cabinets established toll free numbers to answer citizens' questions and offered a voluntary well-testing program for any nearby resident who wished to have their water tested. This was in addition to the radiation monitoring and control program the Commonwealth already had in place outside the facility fence.

Despite the fact that I have seen no imminent threat to public health in the region, my administration's efforts have led me to one obvious and inescapable conclusion. The Paducah Gaseous Diffusion Plant site is one of the most environmentally contaminated in the South, and the federal government is not devoting the necessary funds to meet its obligation to clean it up. And although there is no immediate threat, the nature of the environmental threat is growing and could eventually impact public health.

In 1994, this site was placed on the National Priorities list under the Superfund legislation as one of the most contaminated sites in the country. Pursuant to that designation, our Natural Resources and Environmental Protection Cabinet entered into a tripartite Federal Facilities Agreement with the U.S. Department of Energy (DOE) and the USEPA, whereby the DOE agreed to fund and complete the cleanup of the site by the year 2010. This agreement was finally signed in 1998 and contains significant milestones to be achieved along the path to completion of the cleanup process. We wanted, and felt it would be reasonable to have this work done by 2007. Based on the current rate of progress, it won't be cleaned up in our lifetime.

Working together, these three agencies have identified many areas that must be remediated in order to complete the cleanup. I will not go into detail on all of the items to be addressed, but I feel the subcommittee should hear a little about some of the major concerns.

First, the area known as "Drum Mountain" is a major concern. Drum Mountain (a portion of which is blown up in the photo behind me) is five acres of radioactively contaminated waste materials and scrap metal contained and accumulated on-site since the 1950s. It constitutes a significant environmental hazard because dispersion and surface water runoff contribute to contamination of the area. Moreover, our state agencies suspect that uncharacterized waste materials are disposed of beneath Drum Mountain and that its seepage and these waste materials are contributing to the contamination of the migrating groundwater plumes.

Second, the groundwater plumes, which I just mentioned, are a source of significant concern. These plumes contaminate an underground aquifer of 60–110 feet in depth and are migrating toward the Ohio River in a northwesterly and northeasterly direction. In its recent Phase I investigative report on Paducah, DOE's investigative team admitted that they do not know how far the plume has traveled. Additionally, DOE is having difficulty stopping the advance of the plume with its pumping and treatment because it cannot fully identify the source of the contamination. Our agencies believe that the plumes have reached the Ohio River and are dispersing radioactive Technetium-99 into the river.

Third, surface water runoff and groundwater migration have led to the detectable contamination of Technetium-99, PCBs and trace amounts of transuranics in Little and Big Bayou Creeks, which are tributaries of the Ohio. This contamination must be remediated.

Fourth, the North-South Diversion Ditch, which has tested positive for higher than expected amounts of transuranics such as plutonium and neptunium, must be addressed. At present, we understand that workers at Paducah were not even warned that this ditch was contaminated with transuranics. It sits exposed and is not contained in any manner. This ditch is adjacent to the major buildings on site suspected of transuranic contamination, and in addition to the ditch, these buildings must be cleaned up and decommissioned.

Fifth, all solid and hazardous waste landfills and disposal areas must be identified and characterized by DOE. The Commonwealth has identified to DOE over 200 potential hazardous and solid waste disposal areas on site—about three-fourths of which DOE has failed to fully identify and characterize. In addition, DOE must characterize and remove any radioactive materials contained at two landfills that have been seeping radioactive material.

Sixth, DOE must expeditiously address the drums and cylinders currently in outdoor storage. At present, DOE has over 8000 drums of low-level radioactive waste stored outdoors in containers not designed for long-term storage. Also, although not contained as a milestone in the Federal Facility Agreement, DOE maintains over 37,000 cylinders (over 400,000 metric tons) of depleted Uranium on site. This material must be converted to a more stable form before it is either removed or properly stored.

As you can see from my brief and non-inclusive list of some of the significant environmental hazards, Paducah is a site that demands the immediate attention of the DOE and the federal government.

As I have learned more about the nature of the environmental hazards at the Paducah site, I have become most alarmed not by the extent of the contamination (although it is alarming)—but by the fact that the DOE does not currently have, nor does it plan to request in the near future, sufficient funds to address these environmental concerns.

Mr. Chairman the people of Paducah and Kentucky deserve better than that.

Until very recently, the DOE has estimated that it would take a little over $700 million to complete the cleanup by 2012—and has planned budget requests totaling only $630 million through fiscal year 2010. These projected funding figures anticipated huge funding increases beginning in 2007—despite the fact that the environmental and site management at Paducah has been funded at approximately $38 million per year over the last several years. Moreover, since 1995, environmental funding at Paducah has been steadily declining.

Mr. Chairman and members of the subcommittee, in the Paducah Phase I report released by DOE last week, their investigative team basically admitted two critical facts. First, it admitted that the current cleanup schedule is unrealistic based on current funding levels—and that at current levels, the cleanup cannot be completed before even 2020.

Second, DOE's investigative team admitted DOE's estimated cleanup costs are based on faulty assumptions. In particular, the report reveals that the DOE's future funding numbers are based on proposed savings through recycling of hazardous scrap metal, limiting the number of remedial investigations despite the extent of the problem, capping all waste material found on-site (instead of removing it) and replacing the current pump and treat water remediation with an untested alternative. The report goes on to say that cost savings such as these have never been previously demonstrated.

Mr. Chairman it is time for the Department of Energy to reassess the cost of this cleanup and to be forthcoming about the true projected costs.

When I first began to understand the magnitude of this under-commitment of funds, I asked our cabinets and agencies to independently assess what they believed the cleanup would actually cost. After working through the milestones contained in the agreement, Kentucky State Government now believes that in order to complete the environmental cleanup at Paducah by the year 2010, the cost will be closer to $1.37 billion. If you factor in the funds necessary for the conversion of the depleted uranium in the exposed cylinders and final assessment and management costs, the figure rises to $1.9 billion. I have provided the subcommittee with attachments to my written statement that elaborate upon the Commonwealth's assumptions and that show where we differ from those of DOE.

At present, the DOE environmental and site management activities at the Paducah Gaseous Diffusion Plant are funded from the Uranium Enrichment Decontamination and Decommissioning Fund (D&D Fund) contained with the DOEs Environmental Management Budget. This fund is initially allocated to DOE's Oak Ridge Facility for the cleanup activities at Oak Ridge, Paducah and Portsmouth. Total allocations from this fund for fiscal year 1999 were about $220 million—of which Paducah received about $36 million. We believe this is not a rational division of the $220 million, especially in light of the fact that this $36 million is largely used for ongoing site management activities, with only about $11 million per year going to actual environmental remediation.

Thus, as you can plainly see, DOE last year spent about $11 million on what is approximately a $1.4 billion problem over the next 10 years. Members of the committee, that is not in even in the ballpark of what is necessary.

Moreover, I find it particularly upsetting that the D&D fund takes in about $610 million per year in receipts from both general revenues and a special federal surcharge on power companies that use nuclear fuel. The D&D fund has a positive balance on paper of over $1.5 billion and its excess yearly revenues are used for other budgetary priorities.

Members of the subcommittee, I think that is unfair and I think it breaks a fundamental compact with the communities that have accepted these three facilities. [I am today delivering to the committee a letter signed by Governors Taft, Sundquist and myself asking the U.S. Congress to restore these dedicated receipts to their intended purposes of cleaning up the uranium enrichment sites.]

Mr. Chairman, as I have discussed with you previously, I am asking the U.S. Congress, the DOE, OMB and the White House to dedicate at minimum an additional $100 million per year to the cleanup at Paducah. Only a figure of this magnitude can get us moving toward completing the cleanup by 2010—as the Federal Facilities Agreement mandates. Additionally, I am asking that the DOE not proceed on a milestone by year basis, but that they begin remediating the most pressing priorities immediately and simultaneously. The people of Paducah and Kentucky deserve at least this.

I have already informed the administration that if they are going to be an environmental administration in a regulatory fashion, passing the cost along to the customers of private companies, then they must also be an environmental administration when a matter of federal financial responsibility arises. I have, in the strongest terms, urged the administration to ask for enough money to do this job. If they do, I ask the Congress to approve it. If they don't, I ask the Congress to increase the appropriation sufficiently to do the job. I call upon the Congress to find a way to work with the DOE to fully fund the D&D program for its intended purposes and to make certain that funds are available to complete the cleanup at Paducah by 2010. Completing this obligation of the federal government to the people of Paducah is your responsibility as well. I stand ready to work with our congressional delegation and the political leadership of both parties to help make certain this obligation is met.

As a result of the signing of the Federal Facilities Agreement, the Commonwealth has several legal means at its disposal to enforce this cleanup agreement, including mediation and possible subsequent legal action. If current funding levels are maintained, the Commonwealth believes that DOE will be in default on this agreement as early as fiscal year 2001. Let me assure this subcommittee and the people of Paducah and Kentucky that I will continue my efforts on this issue and that our administration will use every legal means at its disposal to make certain these obligations of the federal government are met. The citizens of Kentucky who have supported this facility for over 45 years deserve no less.

I would like to thank the chairman and members of the subcommittee for the opportunity to appear before you today, and I would be happy to entertain any questions.

Senator MCCONNELL. Governor, we will make that letter from the governors part of the record. We appreciate your coming, and I have a feeling that we are going to be involved in this for a long time to come. This is going to be a long march. I want to thank you for the contribution that you are making at the State level to keeping the heat on, and we will try to do the same up here.

Thank you very much.

The first panel—and I would like to ask the witnesses to try to limit their testimony to about 5 minutes so we have plenty of time for questions, is comprised of David Fuller, who is president of the chemical workers union at the Paducah plant, who will testify on worker radiation exposure; Dr. Steve Markowitz, who is currently performing the health physics study for the Department of Energy and the workers union to evaluate work-related illnesses; and Dr. Richard Bird, who recently completed an exposure study of workers at the Oak Ridge enrichment facility, and reports indicate that many of the tested workers may well have been harmed.

I would like to recognize Mr. Fuller's wife Catherine, who is with us today. Mrs. Fuller, would you please stand up so you can be recognized. Hello. Thank you for joining us.

All right. Well, let us start with David Fuller.

STATEMENT OF DAVID FULLER, PAPER, ALLIED-INDUSTRIAL, CHEMICAL AND ENERGY WORKERS UNION, LOCAL 5–550, PADUCAH, KY

Mr. FULLER. Thank you, Mr. Chairman for the opportunity to come here today and testify before you. My name is David Fuller. I am President of the Paper, Allied-Industrial, Chemical and Energy Union, Local 5–550. PACE represents approximately 850 hourly workers who are employed by USEC at the Paducah gaseous diffusion plant. Our members work in operations, maintenance, and environmental management. 28 hourly workers are slated for transfer to Bechtel-Jacobs in the near term.

I worked at Paducah for 31 years, first as a process operator and later as an electrician. I am a member of the Paducah Site-Specific Advisory Board which advises DOE on its cleanup program. I am also a member of the Paducah Community Re-Use Organization.

I want to make clear at this time that PACE is not a party to any of the litigation that is presently ongoing at Paducah.

NATURE OF WORKER EXPOSURE

First, let me summarize the highlights of testimony offered by our members at previous Congressional hearings in September. For decades workers were not provided respiratory protection while working in the uranium dust, asbestos, and toxic metals. During the processing of irradiated reactor tails into uranium hexafluoride, workers were unknowingly exposed to plutonium, neptunium, and fission products. Until a Washington Post article appeared on August 8, 1999, most workers did not know they were potentially being exposed to plutonium.

The Paducah site did not have a contamination control program for 40 years, leading to the contamination of workers' clothes, shoes, and skin. This resulted in workers tracking contamination off site and into their homes.

Uranium fires self-ignited when dumping uranium chips into open pits. Workers were directed to smother these fires by piling dirt over the burning uranium.

After the site stopped recycling irradiated reactor tails, DOE used the processing building for an employee locker room and a computer repair shop for another 13 years, even though radiation was measured at up to 350,000 disintegrations per minute in locker rooms and 175,000 dpm in showers and toilet areas. These areas should have been posted as contamination areas and not used for purposes that resulted in intimate human contact. DOE enforcement personnel have never set foot at Paducah to investigate the compliance status of the site's radiation protection program.

NEPTUNIUM EXPOSURE

Second, let me summarize the key points that will be in my testimony today. An Atomic Energy Commission memo from 1960 regarding Paducah stated: "There are possibly 300 people at Paducah who should be checked out for neptunium, but they are hesitant to proceed to intensive studies because of the union's use of this as an excuse for hazard pay."

That memo went further and urged Carbide to "get post mortem samples of any of these potentially contaminated men for correlation of tissue content with urine output, but I am afraid the policy at this plant is to be wary of the unions and any unfavorable public relations."

Apparently, management was reluctant to test the deceased for uptakes of neptunium, much less the living. What this memo tells me is that the failure to disclose these hazards to use, to monitor us, was not a happenstance thing. It was a calculated decision. The memo said, if we do the conscionable thing and perform the studies it will cause us discomfort or cost us monetarily.

National security was certainly not the logic for this decision. The AEC faced a simple question: Are we willing to risk lives or pay money? This decision was not a decision made by just any employer. This was a deliberate decision allowed by my government, the institution who is supposed to protect my welfare and to ensure the blessings of liberty to me.

Officials made a cynical choice. The only thing more cynical would be for government to find a way to turn away from this today, now that the facts have come out, and to just do nothing.

MEDICAL AND OTHER HEALTH BENEFITS NEEDED

What we have learned makes us genuinely afraid of what may happen in the future. I personally carry that fear. Medical monitoring by certified occupational physicians is needed today to identify diseases which hopefully can be caught early enough to be successfully treated.

The DOE's medical monitoring program needs to be funded, as promised by the Secretary of Energy. Monitoring is imperative, but without any other remedy, monitoring is simply a process to watch people get sick and eventually die. The workers at Paducah and other sites deserve more than monitoring. They deserve: First, health insurance coverage for all at-risk workers and their spouses through retirement. If we lose our jobs, we will carry the stigma of "glow in the dark" workers, making it almost impossible to find new jobs with health insurance.

Second, coverage for the work force under a Federal workers compensation system that reverses the burden of proof on the government to demonstrate that work place exposures did not lead to illness. With respect to establishing a Federal workers comp program for DOE nuclear workers, Congress has already established a precedent for compensating others who bore the consequences of the nuclear arms race, and they include: The individuals exposed to radioactive fallout downwind from nuclear weapons tests; Marshall Islanders who were exposed to fallout; military personnel participating in weapons testing; civilian weapons test site workers; uranium miners; soldiers guarding the outside of U.S. nuclear weapons production facilities; and, of course, subjects of human radiation experiments.

We are recommending that Congress add coverage of radiogenic cancers to the proposal made by Secretary of Energy Bill Richardson to compensate beryllium disease victims using the Federal Employees Compensation Act as a model. Under this model, a set of presumptions for specific diseases is established. This is essential because DOE currently does not have accurate or complete records of exposures to radioactive substances. Absent this data, the burden of proof upon workers is insurmountable.

This is not about writing a blank check to nuclear workers. What this redresses are the costs which were shifted from the DOE onto the shoulders of its work force, a cost the government never internalized in prosecuting the Cold War.

CLEANUP EFFORTS AT PADUCAH

Allow me to shift focus just for a moment to certain budget inequities affecting Paducah. DOE's budget reveals that $62.5 million, nearly one-third of the D&D budget, is going for removing machinery from three buildings at the Oak Ridge K–25 site, a project which the State of Tennessee declares is not a risk-driven project. By contrast, the entire D&D budget for Paducah is only $37.5 million. How can DOE justify this allocation while at Paducah a plume of contamination is migrating toward the Ohio River at the rate of one foot per day; and nuclear criticality safety concerns are uncharacterized and not being addressed?

Paducah can best rectify the mismanagement and inequities by establishing a Portsmouth-Paducah operations office with its own budget and contracting authority. Paducah's budget is only 3 percent of the entire Oak Ridge budget and Paducah appears to be getting less than 3 percent of Oak Ridge's management's attention. Paducah will continue to suffer as long as we are controlled from a distracted, if not disinterested, field office 350 miles away.

This is the same logic that led Ohio Senators to create the Ohio field office with jurisdiction over Fernald, Ashtabula, Mound, and West Valley.

USEC's future is growing more uncertain and the socioeconomic transitions at Paducah and Portsmouth will eventually include the decontamination and decommissioning of the gaseous diffusion plants. Oak Ridge has not and cannot successfully manage this from 350 miles away.

PREPARED STATEMENT

I thank you very much, Mr. Chairman, and I will be happy to try to answer any questions that I can.

[The statement follows:]

PREPARED STATEMENT OF DAVID FULLER

WHAT IS NEEDED TO PROTECT CURRENT AND FORMER WORKERS FROM PAST AND PRESENT HAZARDS

I am David Fuller, President of the Paper, Allied-Industrial, Chemical & Energy Workers Union, Local 5–550 ("PACE"). PACE represents approximately 850 hourly workers who are employed by USEC at the Paducah Gaseous Diffusion Plant in Paducah, Kentucky. Our members work in operations, maintenance, waste management, environmental restoration, decontamination & decommissioning and escort individuals who lack security clearances. Twenty- eight hourly workers are slated for transfer to Bechtel-Jacobs in the near term.

I have worked at Paducah for 31 years, first as a process operator and later as an electrician. I want to make clear that PACE is not a party to the litigation at Paducah at this time. Today's testimony will focus on:

—Evidence that the Atomic Energy Commission ("AEC") and Union Carbide intentionally kept workers in the dark about their exposures to transuranic elements at Paducah. A 1960 memo explains the government's rationale: fears of adverse publicity and concerns about the "union's use of this as an excuse for hazard pay."

—Secretary Richardson announced a medical monitoring program that would evaluate current workers and accelerate the monitoring of former uranium enrichment workers at Paducah, Portsmouth and Oak Ridge. This initiative has not been funded. Only 350 former Paducah workers will be monitored this year, even though 2,000 current and former could be monitored. The "worker exposure" assessment announced by Secretary Richardson was shut down on October 22 due to lack of funding.

—The government has knowingly placed workers in harm's way, and failed to inform, protect and monitor them. And it did so for economic and public relations reasons, not for reasons of national security. Under these circumstances, where workers can never establish causation, it is necessary for the federal government to establish a federal workers compensation system that cares for those who became ill while serving their country.

—Safety considerations raised by DOE's Phase I Independent Investigation of the Paducah Gaseous Diffusion Plant (October 1999) command re-evaluation of DOE's Management and Integrating contracting strategy—which relies exclusively on performing cleanup with groups of subcontractors. Bechtel is cutting safety oversight staff and complicating the protection of worker safety with the introduction of subcontractor workers with little or no knowledge of site hazards. It may be safer for Bechtel to self-perform cleanup work.

—The DOE's Oak Ridge Operations Office is budgeting more cleanup money for a single non-risk driven project at Oak Ridge than it is providing for the entire Paducah site. Paducah's entire budget is $37.5 million out of a $240 million D&D request for fiscal year 2000, despite uncharacterized criticality risks and a toxic plume migrating towards the Ohio River. Meanwhile, Oak Ridge retains $122 million/year, including $62.5 million/year for a project that Tennessee environmental regulators deemed low risk.

—Oak Ridge Operations Office has been lackadaisical, at best, in the oversight of worker health and safety at the uranium enrichment plants for the past 20 years. Paducah and Portsmouth have been treated like unwanted stepchildren.

—It is time for the creation of a Portsmouth/Paducah Operations Office to manage these two sites. EPA's representative on the Paducah Site Specific Advisory Board concurs. The environmental and safety problems are far too complex to be directed by telephone from 350 miles away in Oak Ridge.

—DOE is proposing to recycle radiologically contaminated metals from the Paducah site to offset cleanup costs. No federal standard exists, and the American public has opposed putting rad metals into products that will come into intimate human contact. The Paducah Site Specific Advisory Board passed a consensus resolution opposing this proposal. Congress needs to assure that the price tag for cleaning up "barrel mountain" at Paducah is not dependent on putting radioactive braces on the teeth of America's children.

I. Summary of PACE testimony on Paducah at previous congressional hearings this year

In testimony before a field hearing of the Senate Energy Committee, Subcommittee on Energy Research and Development, in Paducah, Kentucky on September 20, 1999, and before a hearing of the House Commerce Committee, Subcommittee on Oversight and Investigations, in Washington, DC on September 22, 1999, PACE members described how:

—For decades, workers were not provided respiratory protection while working in uranium dusts, asbestos and toxic metals. During the process of converting reactor tails into uranium hexaflouride—the feed material for the enrichment plant—workers were unknowingly exposed to uranium dust laced with plutonium-239, neptunium-237, and technetium-99. Until a Washington Post article appeared on August 8, 1999, most workers did not know they were potentially exposed to plutonium.

—The Paducah site did not have a contamination control program for 40 years, leading to contamination of workers' clothes, shoes and skin. This led to workers tracking contamination off site and into their homes.

—Uranium fires self-ignited by dumping uranium chips into open pits. Workers were directed to smother these fires by piling dirt over the burning uranium. Uranium self ignited because, in certain forms, it is pyrophoric.

—After the site stopped processing neptunium and plutonium laced reactor tails in the C–410 building, DOE used this building for an employee locker room, electrical maintenance, and a computer repair shop for another 13 years, even though:

  —Radiation was measured at up to 350,000 dpm (disintegrations per minute) fixed contamination in locker rooms. Shower and toilet areas had 175,000 dpm fixed.

  —These areas should have been posted as contamination areas, and not used as a change room.

—DOE enforcement personnel have never investigated the compliance status of the Paducah radiation protection program since the Price Anderson Act enforcement program was initiated in 1996.

—A majority of current and former workers are afraid that may have been exposed to substances like plutonium without proper protection and that they will, as a result, be stricken with a fatal disease. Health insurance and a federal workers compensation system tailored to the unique radiation (and other) hazards is needed to remedy some of the harms from past wrongdoings of the DOE and its contractors.

II. The Atomic Energy Commission and its contractors intentionally kept workers in the dark about exposures to Neptunium-237 a bone seeking radioactive element— at Paducah. The Government's rationale: fears of adverse publicity and concerns about the "union's use of this as an excuse for hazard pay."

A March 11, 1960 memo Neptunium-237 Contamination Problem, Paducah, Kentucky, February 4, 1960, by C.L. Dunham, MD, the Director of the Atomic Energy ("AEC") Commission's Division of Biology and Medicine, and H.D. Bruner, MD, Chief of Medical Research Division of Biology and Medicine, stated:

> There are possibly 300 people at Paducah who should be checked out [for Neptunium], but they are hesitant to proceed to intensive studies because of the union's use of this as an excuse for hazard pay. (Exhibit "A")

Neptunium-237 has a radioactive half-life of 2,140,000 years. Once in the body it concentrates in the bones and liver. With respect to the adequacy of respiratory protection, the memo's authors stated:

> I don't have too much faith in masks, and the dust particles here are about 0.5 micron, the very worst size biologically speaking.

The memo urged Union Carbide to:

> Get post mortem samples on any of these potentially contaminated men for correlation of tissue content with urine output, but I'm afraid the policy at this plant is to be wary of the unions and any unfavorable public relations.

Apparently, management was reluctant to test the deceased for uptakes of neptunium, much less the living. The AEC doctor concluded his memo stating:

> Thus, it appears Paducah has a neptunium problem, but we don't have the data to tell them how serious it is.

What this AEC memo tells me is that the failure to disclose these hazards to us, to monitor us, was not a happenstance thing, it was a calculated decision. The memo said if we do the conscionable thing and perform the studies, it will cause us discomfort or cost us monetarily. National security was not the logic. The AEC and its contractor faced a simple question: are we willing risk lives or pay money. This decision wasn't a decision made by just any employer. This was a deliberate decision allowed by my government, the institution who is supposed to protect my welfare and to ensure the blessings of liberty to me.

Officials made a cynical choice. The only thing more cynical would be for government to find a way to turn away from this today—now that it has come to light and to not step up to the plate and take responsibility.

III. Medical monitoring, health insurance and a federal workers compensation system is needed for those whom the government knowingly placed in harm's way, and failed to inform, protect and monitor.

To learn through a recently released memo of March 1960, that the government made a deliberate decision not to monitor our exposures or show proper concern for our health and safety, has created real anxiety. Our employer provided erroneous re-assurances, not information. Now that facts are coming out, we are genuinely afraid of what may happen in the future. I personally carry that worry.

One consequence of the lack of monitoring is that we have little or no means to prove a worker's compensation claim related to radiation induced illness. The data doesn't exist. Another is that if we lose our jobs at the enrichment plant, we will suffer the stigma of "glow in the dark workers," thus making it almost impossible to find a new job with health insurance.

Medical monitoring by certified occupational physicians is needed today to identify diseases which hopefully can be caught early enough to be successfully treated. The DOE's medical monitoring program needs to be expanded and funded, as promised by the Secretary of Energy, so that any nuclear worker who wants an exam at Portsmouth, Paducah and Oak Ridge can obtain one. Monitoring is imperative, but without any other remedy, monitoring is simply a process to watch people get sick and die.

The workers at Paducah and other sites deserve more than monitoring. They deserve:

(1) Coverage for the workforce under a federal workers compensation system that reverses the burden of proof onto the government to demonstrate that workplace exposures didn't lead to illness, in light of DOE's failure to monitor and adequately protect workers from radiation and other toxic risks.

(2) Health insurance coverage for all at-risk workers and their spouses through retirement. The harm to humans must be treated as seriously as the insult to the environment.

Today, DOE spends nearly $6.0 billion on environmental cleanup and $7.5 million on monitoring at-risk former workers. Resources must be committed to assure equal consideration.

IV. There is ample precedent for the Government to establish a Worker's Compensation System that cares for those who became ill, because the Government failed to disclose and provide protection from hazards connected to nuclear weapons production and testing.

Congress has established the precedent to compensate those who bore the consequences of the nuclear arms race. They include members of the American public exposed to fallout downwind from nuclear weapons tests, Marshall Islanders who were exposed to fallout, military personnel participating in weapons testing, civilian weapons test site workers, uranium miners, and soldiers guarding U.S. nuclear weapons production facilities.

The findings and recommendations of the Presidential Advisory Committee on Human Radiation Experiments ("ACHRE"), built upon these precedents, and established the very important principal of redressing wrongs to people put at risk without their knowledge or consent. The ACHRE Report's "Recommendation 1" states:[1]

> The government deliver a personal, individualized apology and provide financial compensation to the subjects or their next of kin of human radiation experiments in which efforts were made by the government to keep information from these individuals or their families, or from the public, for the purpose of avoiding embarrassment or potential legal liability, or both, and where the secrecy had the effect of denying individuals the opportunity to pursue their personal grievance. (emphasis added)

This recommendation was accepted by the President and has been implemented. In addition to this principle there are several additional principles that provide powerful justification to establish a comprehensive compensation program for DOE nuclear weapons workers across the country.

Since World War II federal nuclear activities have been explicitly recognized by the U.S. Government as a ultra-hazardous activity under law. Nuclear weapons production involved extraordinary dangers, including potential catastrophic nuclear accidents that private insurance carriers would not cover, as well as chronic exposures to radioactive and hazardous substances that, even in small amounts, could cause medical harm. For these reasons, the U.S. Government extended blanket indemnification for its contractors. Under the Price-Anderson Amendments to the 1954 Atomic Energy Act, contractors were held harmless, even for criminal acts or willful negligence.

Since the inception of the nuclear weapons program and for several decades afterwards, large numbers of nuclear weapons workers at DOE sites across the country were deliberately put at excessive risk without their knowledge and consent. In the late 1940's and 1950's, it was brought to the attention of the leadership of the AEC on several occasions that numerous workers were overexposed to federal sites in New Mexico, Washington, New York, Ohio, Colorado and Tennessee.[2] In some instances, workers showed evidence of medical harm.

At Paducah, workers asked for protective clothing in numerous written requests where radiation was likely to get on their clothes. Union Carbide usually denied them. One 1968 grievance by maintenance mechanics, who were overhauling contaminated pumps and valves, stated:

> We ask that we be given this protective clothing [coveralls] back. We further ask that the company be responsible for all hazards and costs from any contamination or radiation carried from this plant into our homes, autos and other areas by the aggrieved employees. (Exhibit "B").

Union Carbide denied this request stating that: "the level of alpha radiation count was not meaningful in itself. It was explained that alpha was injurious only if it was ingested into the body and no provisions for clothing would provide protection for this."

Alpha contamination, such plutonium, could easily be ingested from clothes that had contamination particles on them, or spread onto furniture or food at home. In 1974, our union local finally negotiated the right to protective clothing upon demand. However, coveralls do not constitute an effective contamination control policy.

A 1991 incident in which workers contaminated the plant, their lockers and brought the radiation into their homes underscores both the necessity of a contamination control program that was implemented in 1990, and the enormous hole in the radiation protection program that persisted for nearly 40 years since the Paducah plant opened.[3]

The DOE's practice of misleading workers, either by acts of omission or commission, is a pervasive and well-established government policy. Even into the present time, numerous official reviews and findings reported a continuing problems at DOE sites across the country, where workers were overexposed and not told. In 1951, the AEC's Advisory Committee on Biology and Medicine was told by a federal official that exposures to radiation at AEC plants was "a very serious health problem. This problem is present in other AEC manufacturing plants and will be important in new

---

[1] *Advisory Committee on Human Radiation Experiments, Final Report*, October 1995, U.S. Government Printing Office, pp. 801.

[2] Report of the Majority Staff of the Committee on Governmental Affairs, *Early Health Problems of the Nuclear Weapons Industry and Their Implications for Today*, December 1989, Washington, DC.

[3] *Investigation Report, C–337–A Contamination Incident at the Paducah Gaseous Diffusion Plant on August 23, 1991*, KY/E–112, September, 199.

installations not only from the standpoint of real injury but because of the extreme difficulty of defense in cases of litigation."[4]

The documents uncovered by ACHRE revealed that the suppression of health and safety information was directed not only at nuclear weapons workers and their representatives, but the communities as well. A 1947 memo from the AEC Director of Oak Ridge operations to the AEC General Manager stated:

> Papers referring to levels of soil and water contamination surrounding Atomic Energy Commission installations, idle speculation on future genetic effects of radiation and papers dealing with potential process hazards to employees are definitely prejudicial to the best interests of the government. Every such release is reflected in an increase in insurance claims, increased difficulty in labor relations and adverse public sentiment.[5]

In October 1947 Oak Ridge recommended to AEC Headquarters that the AEC Insurance Branch routinely review declassification decisions for liability concerns:

> Following consultation with the Atomic Energy Commission Insurance Branch, the following declassification criteria appears desirable. If specific locations or activities of the Atomic Energy Commission and/or its contractors are closely associated with statements and information which would invite or tend to encourage claims against the Atomic Energy Commission or its contractors such portions of articles to be published should be reworded or deleted. The effective establishment of this policy necessitates review by the Insurance Branch as well as the Medical Division prior to declassification.

In late 1948 the AEC Declassification Branch found that a study of the effect of gamma radiation on Los Alamos workers' blood could be declassified as it fell within the field of "open research." The AEC Insurance Branch called for "very careful study" before making the report public:

> We can see the possibility of a shattering effect on the morale of the employees if they become aware that there was substantial reasons to question the standards of safety under which they are working. In the hands of labor unions the results of this study would add substance to demands for extra-hazardous pay knowledge of the results of this study might increase the number of claims of occupational injury due to radiation and place a powerful weapon in the hands of a plaintiff's attorney.[6]

As noted above, this same policy was revealed through the March 11, 1960 memo by the AEC biomedical officials where they recognized that "possibly 300 people at Paducah should be checked out" for neptunium contamination, but that there was hesitation to "proceed to intensive studies because of the union's use of this as an excuse for hazard pay."

At the Portsmouth site, Goodyear Atomic issued a *Health Physics Philosophy as a Guide for Housekeeping Problems in the Process Areas,* which it distributed to all supervisors on August 27, 1962. While management assured workers there was no hazard at the uranium enrichment facility in Portsmouth, Ohio, it warned supervisors:

> We don't expect or desire that the philosophy will be openly discussed with bargaining unit employees. Calculations of contamination indices should be handled by the General Foreman and kept as supervisional information in deciding the need for decontamination. (Exhibit "C")

The DOE currently does not have accurate and complete records of exposures to radioactive and hazardous substances—which unfairly places the burden of proof of harm upon workers. According to the DOE's Office of Environment, Safety & Health, from World War II until 1989, radiation doses received from inhalation or ingestion were not estimated or included in worker dose records. Although, DOE took sporadic urine and other samples, contractors made little effort to calculate internal exposures, until they were required by the DOE's Price Anderson Act regulations that became effective in 1996.

---

[4] Atomic Energy Commission, Advisory Committee for Biology and Medicine, Notes, January 12, 1951, U.S. Department of Energy Archives, Germantown, MD.

[5] *Advisory Committee on Human Radiation Experiments, Final Report,* October 1995, U.S. Government Printing Office, pp. 627.

[6] *Advisory Committee on Human Radiation Experiments, Final Report,* October 1995, U.S. Government Printing Office, pp. 656, citing Clyde E. Wilson, Chief of Insurance Branch, to Anthony C. Vallado, Deputy Declassification Officer, Declassification Branch, 20 December 1948.

The reconstruction of individual worker doses is extremely costly and fraught with uncertainties and error. Earlier this year, the Director of DOE's Office of Enforcement conducted a survey which found that many DOE's contractors were not properly monitoring internal ingestion of radiation doses. A July 15, 1999 memo stated:[7]

Evaluation and assignment of worker doses are consequently, inadequately and/or inaccurately performed such that compliance with annual DOE limits for personnel exposure may not be assured. (emphasis added)

The deficiencies found in 1999 include: failure to advise workers of their doses; failure to analyze for all radionuclides to which workers were exposed; dose assessment for workers that have an uptake were not completed; internal dose assessments are not accurate; failure to perform in vivo bioassays; and rad worker restrictions are not implemented in a timely manner.

Since World War II the DOE and its predecessors have been self-regulating with respect to nuclear safety, and occupational, safety, and health. DOE relies on contractors to perform about 90 percent of its work, including the day-to-day operational responsibility to guarantee a safe working environment. For the past 20-years, DOE's self regulation has been subject of a considerable amount of criticism due to its ineffectiveness.

The DOE's indemnification policies place the full resources of the U.S. Treasury at the disposal of contractors to fight workers compensation claims. Blanket reimbursement of contractors for legal costs is a powerful weapon to prevent workers or their survivors from gaining compensation for latent diseases. Secretary of Energy Bill Richardson has conceded that DOE has deployed its full resources to fight workers' compensation claims for occupational diseases, regardless of merit.

And DOE has gone to unlawful extremes to prevent workers from getting compensation. In 1984, a Court of Appeals ruled that the state workers compensation program for DOE contractor employees in Nevada was invalid.[8] The state had a secret agreement with the DOE and its predecessors since the early 1950's which allowed DOE to decide on radiation compensation claims filed by test site workers or their survivors.

DOE had a powerful weapon at its disposal: the AEC (and later on DOE) would reimburse the Nevada Industrial Commission only if the AEC agreed that a claimant's award was justified. If it disagreed, the Atomic Energy Commission and the Nevada Industrial Commission could submit the dispute to arbitration. If the arbitrator ruled that reimbursement is required, the agreement permits the Atomic Energy Commission to seek a de novo determination in a court of law. With that weapon at its disposal, workers were helpless to prevail, until the Nevada compensation system was unmasked and declared unlawful.

The department's handling of the Kentucky worker's compensation claim on behalf of my co-worker, Joe Harding, who was employed at DOE's Paducah facility, is another case in point. Joe died in 1980 from cancer and his wife Clara filed a compensation claim with the Commonwealth of Kentucky on March 1, 1983. She had her husband's bones exhumed, and uranium was found in bone tissue. Dr. Carl Johnson, an expert who analyzed the independent laboratory results, calculated that Joe had 1,700 to 34,000 times normal uranium levels in his bones at the time he left the plant, with a dose of 30 to 600 rem to the bone tissue. Annual worker whole body dose limit is 5 rem/year. The DOE, and its contractor, Union Carbide, opposed this case for some 14 years. Eventually Mrs. Harding settled with Carbide and its insurer for $12,500 in September, 1997.

Several epidemiological studies have shown that DOE workers are experiencing greater than expected risks from dying from certain cancers and other diseases. Over the past 20-years, several studies have shown increased risks of cancer and other diseases among DOE workers. They include workers at Hanford, Rocky Flats, Oak Ridge, Fernald, the Savannah River, a uranium processing facility in upstate New York, and the Santa Susanna facility in California. No such study has been done at Paducah.

We recommend the Congress should establish a federal employee compensation system that redresses the government's failure to protect its workforce. What we propose is the continuation of a 20 year precedent: to provide compensation for those people who were put at risk without their knowledge and consent; who were deliberately misled; who, in some cases, were intimidated by formidable legal resources

---

[7] Memorandum for DOE PAAA Coordinators and Contractors PAAA Coordinators from R. Keith Christopher, *Compilation of Bioassay Issues Reported During the 120 Day Suspension of PAAA Enforcement Actions Related to Internal Dose Evaluation Programs in the Department of Energy Complex*, July 15, 1999.

[8] *Keith L. Prescott v. United States*, 731 F.2d 1388, 1984 (9th Circuit).

of the U.S. government; and who now suffer the consequences. We are proposing to add coverage of radiogenic cancers to the proposal made by Secretary of Energy Bill Richardson to compensate beryllium disease victims using—the Federal Employees Compensation Act as a model. This is not writing a "blank check" to nuclear workers. What this redresses are the costs which were shifted from the DOE on to the shoulders of its workforce a cost the government never internalized in prosecuting the cold war.

It's time that the government assume those costs that are being borne by those who with dedication provided for our nation's defense during some of its darkest hours. In the aftermath of the Cold War, the DOE must make peace with the people who helped this nation prevail.

V. Doe's management & integrator contracting strategy—which relies on performing cleanup exclusively through subcontractors increases health and safety risks by bringing in workers without knowledge of site hazards. The M&I contracting approach requires additional health and safety oversight by the DOE and Bechtel Jacobs.

A. DOE's subcontracting approach increases health and safety risks, and could lead to loss of experienced workers with valuable institutional memory.

The DOE's Phase I Independent Investigation of the Paducah Gaseous Diffusion Plant (October, 1999) by the Office of Environment, Safety and Health states:

> [u]nder the management and integrating contractor concept, a large fraction of the potentially hazardous work will be performed by subcontractor employees, some of whom do not have long-term knowledge of site hazards or controls. (pp.4)

—Bechtel Jacobs subcontractors do not consistently follow safety and health procedures.

—Some recent subcontractor work activities have resulted in unsafe work practices.

—The investigation team observed subcontractor ES&H performance that did not meet DOE requirements.

—There is little oversight of training programs by DOE, and there are no mechanisms to ensure that the training that is provided is adequate.

—DOE has not conducted effective oversight of ES&H or ensured that Bechtel Jacobs and its subcontractors effectively implement all DOE and regulatory requirements.

The experience gap cited by DOE can be solved by using Paducah's incumbent hourly workforce. These workers have the unique site-specific knowledge DOE suggests is needed for the cleanup at Paducah and Portsmouth. That knowledge can save also DOE money on characterization studies, and help prevent the kind of mistakes that surfaced at Pit-9 in Idaho.

For example, Chris Naas, a heavy equipment operator at Paducah for 25 years, testified before the Senate Energy Committee Field Hearing how he was directed to place barrels of waste in the "404 holding pond" that contained "nickel stripper, trichlorethylene, green salt and yellow cake powder." These buried drums are one of the sources of groundwater contamination.

To date, Bechtel Jacobs has refused to commit to use on-site workers for the biggest hazards at Paducah: groundwater remediation, cleaning up "barrel mountain", and decontaminating and decommissioning the empty process buildings. Bechtel will only commit to retaining 28 incumbent hourly workers, which they will flow down to subcontractors for waste management, maintenance, and DUF6 cylinder hauling.

We have asked DOE and Bechtel Jacobs to create a bridge so hourly workers can move seamlessly from USEC to Bechtel Jacobs over the life of the cleanup. We expect that a number of site workers could be laid off as early as July, 2000 when the USEC-Treasury Department Agreement is set to expire. Layoffs could follow in future years as well, as 47 percent of USEC's production has been displaced by imports under the U.S. Russia HEU Agreement.

We have conveyed our concerns directly to the Secretary of Energy, as well as the Assistant Secretary for Environmental Management, the Oak Ridge Operation Office Manager, the Director of the DOE's Office of Worker and Community Transition, and two Senior Policy Advisors to the Secretary of Energy.

Unless there is a change in the thinking at the DOE's Oak Ridge Operations Office and at HQ, no more than a handful of hourly employees with knowledge of site hazards will be retained for the cleanup mission at Paducah or Portsmouth. What we are proposing makes good policy sense and is the right thing to do for the workers. Continued inaction seems inexcusable.

B. Safeguards, such as oversight, are not in place to protect worker safety.

The Phase I Independent Investigation of the Paducah Gaseous Diffusion Plant states:

> [e]xpanding reliance on subcontractors for cleanup and waste management activities will require significantly more surveillance and oversight by both Bechtel Jacobs and DOE personnel who are knowledgeable of DOE requirements. In some cases, these requirements may be more stringent than the subcontractors' normally accepted practices. It has been demonstrated throughout the DOE complex that more active oversight and surveillance at the activity level is necessary to raise the threshold of acceptability for safe work practices and environmental conditions. If DOE is successful in obtaining funding to accelerate cleanup activities at PGDP, significantly more effort must be expended on surveillance and oversight to achieve and maintain the requisite standards for protecting the environment, the public and especially the workers. (pp. 48)

The Phase I Independent Investigation noted that Bechtel Jacobs is planning staff reductions that will further reduce its technical capacity to conduct oversight and surveillance of subcontractor activities. (pp.48) Yet, even after Bechtel Jacobs was briefed on the Phase I Independent Investigation concerns about subcontractor oversight, they nonetheless went forward and awarded a major subcontract at Paducah for waste management to Weskem. This subcontractor is scheduled to commence work in December.

The rush to issue subcontracts before adequate safety mechanisms are in place is driven by the contractor's incentive fee awards. Bechtel Jacobs has $2 million in award fees tied to initiating subcontracting of all workscope by September 30, 1999 (Exhibit "D"). Bechtel Jacobs also has $2 million in award fees tied to a reduction in total headcounts by September 30, 2000. (Exhibit "E"). DOE seems unaware that its performance-based incentive fees have pitted perceived cost-cutting measures against protecting worker health and safety. DOE champions this approach as an example of "contract reform." Down in Paducah, we can stand only so much "reinventing government."

DOE needs to postpone the deadlines facing Bechtel Jacobs to commence subcontracting, at least until adequate safety measures are in place and are validated by the Oversight Team. DOE needs to immediately scrap award fees tied to goals that are jeopardizing safety. These incentive awards are leading to mismanagement at the ground level in Paducah. DOE senior management also needs to re-examine the wisdom of its exclusive reliance on subcontracting. Self-performance by Bechtel Jacobs may be a preferable option.

VI. DOE's Oak Ridge Office has misdirected cleanup funds into non-risk driven projects at the Oak Ridge K–25 site, while ignoring extremely high risk hazards at Paducah.

For fiscal year 2000, funding for cleanup at the three gaseous diffusion sites comes primarily from the Uranium Enrichment Decontamination and Decommissioning Fund, and is broken out as follows:

Fiscal Year 2000 Budget Request (D&D Fund)

Oak Ridge	$122,068,000
Paducah	37,500,000
Portsmouth	37,500,000
Oak Ridge "Off Site"	8,030,000
Oak Ridge Operations	5,100,000
Total	[1] 210,198,000

[1] This excludes $30 million for thorium tails.

DOE's budget reveals that $62.5 million, nearly one-third of the D&D budget, is going for removing machinery from three buildings at the Oak Ridge K–25 site a project which the State of Tennessee declared is NOT a risk-driven project. By contrast, the entire D&D budget for Paducah is only $37.5 million. How can DOE justify this allocation, while at Paducah a plume of contamination is migrating towards the Ohio River at the rate of 1 foot per day, and nuclear criticality safety concerns in 11 DOE material storage areas go uncharacterized?

DOE's motivation to pursue environmentally insignificant projects at Oak Ridge ahead of higher priorities is described in an October 30, 1996 "Project Managers Meeting Notes," which included DOE-Oak Ridge, Tennessee Department of Environmental Conservation ("TDEC") and the EPA. The memo states ("Exhibit "F"):

—"According to DOE, this [K–29/K–31/K–33 Buildings D&D] effort is important not primarily from a risk reduction aspect, but it is important because it is the first large effort by DOE to D&D gaseous diffusion facilities and it will serve as the national precedent for how other similar facilities will be addressed in the future."

—TDEC questioned whether "yet another project was being introduced into the Oak Ridge operations that would be competing for ER (Environmental Restoration) funds."

—TDEC "expressed a reluctance to agree to the [D&D of the 3] buildings as FFA (Federal Facility Agreement) milestones, since risk reduction is not the primary issue."

—"TDEC remarked again, that the K–29/K–31/K–33 D&D effort was not going to be done as a risk reduction priority."

Notwithstanding these reservations, TDEC and EPA acceded to DOE's request, without so much as a public hearing. EPA, as the federal regulator over both Oak Ridge and Paducah, inexplicably allowed this gross misdirection of scarce resources.

Due to the propensity of Oak Ridge to beggar the Paducah and Portsmouth sites for its competing goals, funding requests for Paducah have been declining when environmental risks and regulatory requirements for cleanup are increasing. The Phase I Independent Investigation of the Paducah Gaseous Diffusion Plant noted:

—A 1998 Report to Congress on the use of the decontamination and decommissioning fund did not identify the need for additional funds to keep the contamination at Paducah from spreading to surrounding environment.

—This Oak Ridge-prepared report emphasized accomplishments, but did not discuss challenges faced at the site to reduce and prevent spread of contamination to the environment within a declining budget.

VII.Oak Ridge Operations Office has a history of lackadaisical oversight of worker health and safety at Portsmouth and Paducah

Since NRC and OSHA have no authority over the DOE-controlled areas of the Paducah plant, we depend on the DOE's Paducah Area Office and DOE's Oak Ridge Field Office to police its contractors safety practices. However, Oak Ridge has largely functioned as an absentee landlord, allowing our site's safety profile to deteriorate except when the GAO, Tiger Team Reports or the EH-Oversight Team reveal embarrassing failures.

A July 1980 Comptroller General report, *Department of Energy's Safety and Health Program for Enrichment Plant Workers Is Not Adequately Implemented* (EMD–80–78), found that DOE's Oak Ridge Office had not conducted a safety inspection at any of the three gaseous diffusion plants in Oak Ridge, Portsmouth or Paducah for 2 years and was not adequately responding to worker safety complaints. Unannounced safety inspections were supposed to occur annually at each plant, but even when they were inspected, the Oak Ridge Office "does not, as part of an inspection or any other visit to an enrichment plant, monitor for radiological contamination." Prior to 1980, the report noted that the previous inspection at Paducah was in 1978 and the one before that was in 1976. Oak Ridge explained the absence of inspections on a staff shortage, which the Comptroller General noted was attributable to Oak Ridge paying safety inspectors at a lower grade than elsewhere in the DOE complex.

In 1990, the Tiger Team found a lack of contractor compliance with DOE Orders and mandatory standards in many areas, including worker safety, quality assurance, radiological protection, and control of administrative documents. A survey plan was formulated for transuranics after technetium-99 was found in an off site well. The Tiger Team report noted that DOE was not performing effective oversight to ensure that ES&H initiatives were being implemented.

The Paducah site office was increased from 5 to 12 members after the Tiger Team report in 1990. In 1993, two Site Safety Representatives were assigned to Paducah, primarily to oversee the Congressionally mandated transition to external NRC regulation as part of the creation of USEC as a government-owned corporation. However, by 1997 the Site Safety Representatives were released to other jobs, as the transition to the NRC oversight of the enrichment plant was competed.

Nine years later after the Tiger Team report, a series of Washington Post headlines prompted DOE to initiate another investigation at Paducah. The Phase I Independent Investigation identified numerous deficiencies in the oversight of Bechtel Jacobs and its subcontractors by the Oak Ridge Office.

VIII. Portsmouth/Paducah Operations Office should be created to take charge of these two major sites.

It is time to separate Paducah and Portsmouth from Oak Ridge. The environmental and safety problems at Paducah are too large and too complex to be managed by telephone from 350 miles away in Oak Ridge.

While the new manager assigned to Oak Ridge brings impressive credentials to the job, she cannot overcome the fact that the central focus of her mission is centered on the $1.5 billion that is spent at Oak Ridge: the Y–12 defense facility, Oak Ridge National Labs, the K–25 site. Paducah's budget is 3 percent of Oak Ridge's annual budget. The Paducah and Portsmouth sites are satellite operations. A fully staffed Operations Office with budget and contracting authority that is centrally focused on the problems at Paducah and Portsmouth is a part of the solution. The incremental cost of a new Operations Office will be more than offset by making these sites a high priority instead of an afterthought. This is the same logic that led Ohio Senators to create an Ohio Field Office with jurisdiction over Fernald, Ashtabula, Mound and West Valley.

DOE is planning the construction of two depleted uranium hexaflouride ("DUF6") conversion plants at Portsmouth and Paducah (pursuant to Public Law 105–204). DOE will convert "tails" for the next 20 years, budget permitting. But success for this huge project requires dedicated management focus. Likewise, controlling the source of groundwater contamination, addressing criticality concerns, and resolving the fate of 60,000 tons of radioactively contaminated scrap metal are massive problems that require the full time focus of an Operations Office.

USEC's future is growing more uncertain, and the socioeconomic transitions at Paducah and Portsmouth could be dramatic. Ultimately, the future of Portsmouth and Paducah—hopefully far in the future—will include the decontamination and decommissioning of the gaseous diffusion plants. Oak Ridge has not and cannot possibly manage the myriad of interfaces from 350 miles away.

IX. DOE is proposing to recycle radiologically contaminated metals from the Paducah site to offset cleanup costs. No Federal safety standard exists, and the public has opposes putting radioactive metals into products that will come into intimate human contact.

DOE has issued a draft plan to cleanup the 60,000 tons mountain of radioactively contaminated scrap metals at Paducah which are leaching radiation into groundwater. DOE recommends that Paducah sell the radioactively contaminated nickel, steel and copper to scrap metal dealers as a way to offset the cost of cleanup. These metals would find their way into intimate human contact, such as kitchenware, zippers, baby carriages, orthodontic braces, iron tonics and eyeglasses.

Putting radioactive metals into commerce has generated strong opposition from the steel industry, the scrap metal dealers, the Steelworkers Union and public interest groups. The copper, brass and nickel industries are also raising questions.

There are 9,350 tons of nickel ingots that are contaminated throughout with uranium, technetium-99, neptunium and plutonium. There are no federal standards governing the free release of this metal into unrestricted consumer goods, and impossible technical hurdles to overcome in monitoring the so-called "volumetrically contaminated" metals. DOE has said it will carefully monitor every centimeter of metals it releases. At a time when DOE concedes that radiation monitoring for its workers is deficient, is it believable they will do a 100 percent job monitoring the mountains of scrap metals?

The Paducah Site Specific Advisory Board reviewed the DOE's plan and adopted a consensus recommendation at the August 1999 meeting that opposes the unrestricted and/or free release of this metal into commerce absent a federal standard.

Instead of burying this rad metal, some could be recycled for "restricted" use in nuclear facilities. Congress needs to set rules for DOE, by prohibiting this metal from finding its way into forks and knives that wind up on our dinner tables. Congress needs to assure that the price tag for cleaning up "barrel mountain" at Paducah is not dependent on putting radioactive braces on the teeth of America's children.

X. Conclusion

Workers at Paducah are afraid of what may happen to them in the future, as they have unknowingly worked with radioactive and toxic substances, such as plutonium, that have long latency periods and can have catastrophic results. These workers— who served our nation as veterans of the Cold War production era must not be forgotten.

Medical monitoring is necessary, but insufficient. Workers need health insurance that will be with them throughout retirement. We need a federal workers compensation system modeled after the Federal Employees Compensation Act—that will take care of those of us who are never going to be able to prove our illnesses were work related because the government's conscious decision not to monitor them or advise them of their risks to transuranics.

A Paducah/Portsmouth Operations Office is needed to bring focus to the large challenges faced by these two sites. Oak Ridge cannot manage these complex sites by telephone from 350 miles away.

Congress needs to prohibit the unrestricted release of radioactive metals into everyday commerce.

DOE needs to re-evaluate whether subcontracting is the best means to safely accomplish cleanup at Paducah. At a minimum, DOE needs to direct Bechtel Jacobs to utilize the institutional memory and site-specific knowledge possessed by the incumbent hourly workforce as it executes the cleanup of Paducah and Portsmouth.

STATEMENT OF STEVEN B. MARKOWITZ, M.D., PROFESSOR, CENTER FOR THE BIOLOGY OF NATURAL SYSTEMS, QUEENS COLLEGE, NEW YORK, NY, AND ADJUNCT PROFESSOR, MOUNT SINAI SCHOOL OF MEDICINE, NEW YORK, NY

Senator MCCONNELL. Thank you, Mr. Fuller.

I think what we will do is go on and take Dr. Markowitz's statement and Dr. Bird's and then we will ask questions of all three of you.

Dr. Markowitz is a physician who specializes in occupational and environmental medicine and is Professor of Earth and Environmental Sciences at City University in New York. He is also the director of the worker health protection program and national medical screening program for the early detection of occupational diseases experienced by workers what were formerly employed in nuclear weapons production in various DOE facilities.

He is currently running the workers health testing program at the three gaseous diffusion plants, and he is here today to update us on the progress of the program and the ways it might be expanded and improved.

Dr. Markowitz, we welcome you here, and if you could try to complete your testimony in around 5 minutes that would be great.

Dr. MARKOWITZ. Thank you, Mr. Chairman.

My name is Steven Markowitz and I am an occupational medicine physician, which means I deal with problems that arise in the work place and exposures that impact on health. This is a relatively little known specialty throughout the United States, but nonetheless a very important area. I serve as Professor at Queens College in New York and also Adjunct Professor at Mount Sinai School of Medicine.

My written testimony is longer than what I will speak about today, given the 5-minute limitation, and there are some problems that I discuss that can be read about later. But let me focus on our workers health protection program.

This is a program which is a collaboration between Queens College, PACE International Union, and the University of Massachusetts at Lowell. It was established by the Department of Energy 3 years ago under contract to them. It was initiated under order from Congress, section 3162 of the 1993 National Reauthorization Defense Act. Section 3162 simply said to the Department of Energy, if you can locate former DOE workers at significant risk for occupational disease because they have had untoward exposures in the plants, then they should be medically screened and monitored. We have undertaken such a program at the three gaseous diffusion plants and also at Idaho National Laboratory.

MEDICAL SCREENING AND EDUCATION PROGRAM

The goal of this program is the early identification of work-related conditions at a point at which we can intervene and actually do some good for people. This program is about clinical service. It is not about research. It is about benefiting people who, as you have said, have been in harm's way and now at a relatively late date actually providing something that can be medically useful to them.

In this medical screening and education program, we invite former workers in for screening. In fact, in the last 6 months that we have operated—we began screening about 6 months ago—we have not had to invite people, because we held press conferences at the three sites and we have had over a thousand calls to our toll-free number. There is a great deal of interest in this program, in this type of activity.

We send people to local clinical facilities at the sites under contract to us, where they undergo a medical screening. We also conduct 2-hour workshops conducted by current and former workers, partly under the direction of David Fuller, who spoke previously.

Our goal really is to help people understand, retirees from the gaseous diffusion plants, to help them understand what has happened to them, what kind of exposures they have had. There are uncertainties about those exposures, but we owe them the truth at least about those uncertainties and about what we know. Also, we try to tell people how their health has been affected by working at DOE.

MEDICAL SCREENING EFFORT

It has been a good program to date. We have funds from the Department of Energy to screen 1,200 people this year at all three plants. That is 400 at each site. We have screened about 450. We have conducted education for a little over 400 people. The program has gone well.

We have found—we have not really aggregated our results. We have found some, albeit limited, amount of occupational illness, including asbestos-related disease, emphysema, and hearing loss at these three facilities.

Let me talk about how we can expand this program and make it a lot better. We really have an outstanding opportunity now to alter the program to both expand the number of people who could benefit from our service and also to include lung cancer screening. As you mentioned before, there are about 15,000 or more former gaseous diffusion plant workers who could benefit from this program and there are about 5,000 or so current workers who could benefit from this program. So we would like to expand and conduct medical screening at a faster rate so that we can actually get through all first screening of the people within a limited number of years.

LUNG CANCER SCREENING

I want to focus in on lung cancer screening. This is an issue that has not fully impacted public consciousness yet, but we are really on the threshold of a major advance in screening for cancer. Lung cancer is the number one cause of cancer death in the United States. About 158,000 people this year in this country will die from lung cancer. The death rate is 90 percent of those who get lung cancer, those will die; 90 percent will die from lung cancer.

Despite advances in cancer screening otherwise, for instance in breast cancer, prostate, cervical cancer, colon cancer, lung cancer has remained as the single most common cancer for which screening has not been effective. There now is an effective method for screening for lung cancer. This work was done originally in Japan several years ago and now confirmed in the United States. It was published 3 months ago in Lancet, which is a leading medical journal, a study by Dr. Claudia Henschke and others at Cornell and NYU University Medical Centers in New York.

CT SCANNING FOR LUNG CANCER

Let me just give you the numbers that they looked at. They enrolled 1,000 people, all of whom were smokers or former smokers, in the study and they conducted low-dose CT scan of the chest for those thousand people. They were age 60 or over, and they were both men and women, but otherwise not at excess risk for lung cancer except for the fact of cigarette smoking.

Of those thousand people, they found that 27 had lung cancer. These 27 had nodules in the lung that were cancerous, of those 27, 23 or 85 percent were at the earliest stage of lung cancer. That is to say stage one lung cancer, which is the earliest stage, is eminently curable. Stage one lung cancer, which only appears in a few of the people who present currently, can be cured; 5-year survival for stage one lung cancer is about 80 percent.

CT scanning provides the method of detecting lung cancer at stage one, at the earliest stage. Of those 27 people with lung cancer, 26 received surgery and virtually all of them can be expected to be cured of their lung cancer. So we can deliver with CT scanning a 70 to 80 percent 5-year survival for lung cancer, compared to the current 10 percent 5-year survival for lung cancer. This can be done, and it is much better than the current use of a chest X-ray.

In our screening program so far we have used the chest X-ray because that is all that has been available and because CT scanning is more expensive. We would like now to apply this new technique of CT scanning to detect lung cancer early in the gaseous diffusion plant workers.

Now, why these workers? Well, many of them smoke, so many of them are at risk for lung cancer as a result there. But in addition, many have been exposed to lung carcinogens or cancer-causing chemicals in the workplace, specifically asbestos, specifically beryllium, silica, and now plutonium and neptunium. These are lung carcinogens. They cause lung cancer among humans.

We would like to introduce CT scanning in Paducah, in Portsmouth, and Oak Ridge. Medical advances typically bear fruit in metropolitan areas first. There is a great deal of excitement in the major medical centers in New York about CT scanning, and I am sure in San Francisco and Chicago and the other major cities this kind of work will be implemented for the early detection of lung cancer. But normally in a place like Paducah, a small city like Portsmouth or Oak Ridge, this kind of medical advance will take 3, 5, or 7 years to arrive.

IMPLEMENTATION OF CT SCANNING

With this program, with some funding from Congress, from the Department of Energy, we can implement CT scanning of the lung for early detection among these workers. In fact, it will cost, we estimate—and we have given all the details both to the Department of Energy and to your staff—that in the next 12 months $5.8 million can be spent obtaining a CT scan, putting it on a mobile unit— I have a picture here of what such a unit would look like. We will transport the unit between Portsmouth, Paducah, and Oak Ridge, and employ it full-time, providing CT scanning of the chest for 2,000 former and current gaseous diffusion plant workers.

Of those 2,000 workers, we can expect that we will detect——

Senator MCCONNELL. How long would it take you to do the 2,000 workers?

Dr. MARKOWITZ. 2,000 is what we expect to do in 12 months. Among those 2,000, we can expect to detect several dozen people with lung cancer, most of whom will have early stage lung cancer and can be cured of the disease.

I see my time is up, so let me just stop here and answer any questions later if you would like.

[The statement follows:]

PREPARED STATEMENT OF STEVEN B. MARKOWITZ

My name is Steven Markowitz, MD. I am a physician specializing in occupational medicine, that is, identifying and reducing workplace exposures that impair or threaten human health. After receiving my undergraduate degree from Yale and my medical degree from Colombia University, I completed five years of training in internal medicine and occupational medicine in New York City and had the excellent fortune of training under the late Dr. Irving Selikoff, the noted asbestos researcher at Mount Sinai School of Medicine. I currently serve as Professor at the Center for the Biology of Natural Systems of Queens College and Adjunct Professor of Mount Sinai School of Medicine, both in New York City.

My research interests center on the surveillance and identification of occupational disease. I recently completed a study commissioned by the National Institute for Occupational Safety and Health concerning the extent and costs of occupational disease and injury in the United States (Attachment A).

I wish today briefly to highlight two core problems in occupational health at the gaseous diffusion plants of the Department of Energy, at Paducah, KY; Portsmouth, OH; and Oak Ridge, TN and to discuss our response to those problems through the initiation of the Worker Health Protection Program. I will start first with our response and then briefly elucidate the core problems.

THE WORKER HEALTH PROTECTION PROGRAM

In 1996, we initiated the Worker Health Protection Program (WHPP) at the three Department of Energy gaseous diffusion plants. It is a medical screening and education program established as a collaboration between Queens College, the PACE International Union and the University of Massachusetts at Lowell and with the full cooperation of the employers at the plants. This program developed as a result of Congressional passage of Section 3162 of the National Reauthorization Defense Act of 1993, requiring that the Department of Energy initiate a medical surveillance program for former DOE workers who (a) were at significant risk for work-related illness as a result of prior occupational exposures at DOE facilities, and (b) would benefit from early medical intervention to alter the course of those work-related illnesses. We received a contract from the DOE through a competitive, merit-based review process and have now, after a careful needs assessment and planning process, instituted the Worker Health Protection Program at the three gaseous diffusion plants in Paducah, Portsmouth, and Oak Ridge as well as the Idaho National Engineering and Environmental Laboratory in Idaho Falls.

The goal of the Worker Health Protection Program is to detect work-related illness at an early stage when medical intervention can be helpful. At a broader level, the goal of our program is to help former DOE workers understand whether they have had exposures in the past that might threaten their health and to ascertain whether, in fact, an injury has resulted from these exposures. For the first time, former workers of the DOE gaseous diffusion plants have the opportunity to obtain an independent, objective assessment of their health in relation to their prior workplace exposures by a physician who is expert in occupational medicine. We screen for chronic lung diseases, such as asbestosis and emphysema, hearing loss, and kidney and liver disease. We have not heretofore emphasized cancer screening, because the screening tests available to date for the cancers of concern have been inadequate, and because the gaseous diffusion plants have not historically been considered sites of high radiation exposure. We implement the program through local clinical facilities based a common medical protocol. This is not a research activity, but a clinical service program, intended to be of direct and immediate benefit to participants.

We also provide a two hour educational workshop during which former DOE workers have the opportunity to learn about their past exposures and what they might mean in terms of health. These workshops are run by current and former workers, because they have credibility and expertise. We also believe that a participatory model of education is in and of itself health-promoting.

Our program is highly successful. In the past 5 months, approximately 1,000 former gaseous diffusion plants workers have called our national toll-free number requesting screening appointments. We have screened 450 people and educated 420 people to date. It is early in the project to aggregate results, especially since the first screening participants are a self-selected group and may not reflect the broader experience of the former DOE workforce. We have seen some, albeit limited, work-related illnesses among the screeners to date. As importantly, we have seen a high degree of interest, enthusiasm, and satisfaction with the program.

The Worker Health Protection Program is, however, severely limited by available funding. The DOE provides sufficient funds to screen 1,200 former gaseous diffusion workers per year. Since we estimate that there are at least 15,000 living former GDP workers who are eligible for our program, we will need over 12 years at the current rate of funding to screen each person one time. Clearly, this is inadequate and undermines the intent of Section 3162.

ENHANCING THE WORKER HEALTH PROTECTION PROGRAM

Due to the recently acquired knowledge that gaseous diffusion plant workers have been exposed to transuranic materials and the likely heightened health risks associated with these exposures, we now propose to rapidly expand our medical testing program. We have made this proposal at the invitation of the Department of Energy.

Three significant improvements in the Worker Health Protection Program are worthy of support, as follows:

1. Adding current workers to the screening and education program.
2. Accelerating the pace of testing from 1,200 to 5,750 workers per year.
3. Initiating screening for the early detection of lung cancer through the use of a low-dose computerized tomography (CT) scanning protocol.

We describe herein the rationale and numeric estimates of eligible workers that underlie these three proposed additions to the current program. We also provide some insight into the ability of an accelerated program to meet the needs of workers, both current and former, at these three facilities in the coming years.

Adding Current Workers

Workers presently employed at the three gaseous diffusion plants do not currently receive the benefits of a medical screening and education program that is (a) specifically designed for early detection of work-related disease, and (b) provided by independent, credible physicians and other professionals with expertise in occupational medicine. They do not universally have access to such a program. Yet they clearly deserve it, based on their many years of service to the nation and the occupational risks that they have encountered during this service.

We estimate that the numbers of current workers at the gaseous diffusion plants are: 1,800 at Paducah; 2,000 at Portsmouth; and 1,700 at Oak Ridge K–25 (Table 1). During the next 12 months, we propose screening one-half of current workers, or 900 at Paducah; 1,000 at Portsmouth; and 875 at Oak Ridge K–25. This totals 2,750. Workers with the longest duration at the plant (especially from the mid-1950's to the mid-1970's), or who are deemed to have worked in the highest risk areas will be offered screening first. This program capacity will allow all current workers to be screened within two years. In fact, since not every current worker will wish to participate in the program, all interested current workers will be screened in less than two years.

Accelerating the Medical Screening of Former Workers

The Worker Health Protection Program now screens former gaseous diffusion plant workers at the rate of 400 per year per plant. This pace is constrained only by budget limitations. The estimated number of former workers at the three sites, over 15,000 (7,000+ at Oak Ridge K–25; 5,000+ at Portsmouth; and 3000+ at Paducah), is quite high, indeed much higher than the number of current workers. The above-proposed screening rate for current workers will outstrip the present rate for screening former workers. This is inequitable and contrary to our knowledge of risk, since former workers are at no less risk than are current workers for work-related health problems from having worked at gaseous diffusion plants. We therefore propose to speed up the rate of screening former workers to 1,000 per year at each of the three sites. This totals 3,000 workers per year (Table 1). Since we are currently budgeted to screen 400 per year per site, the requested funds will allow screening of 1,800 additional former workers in the next 12 months. This accelerated screening capacity will enable a higher proportion of former workers to be screened within a limited number of years.

TABLE 1.—ESTIMATED NUMBERS OF CURRENT AND FORMER WORKERS AT GASEOUS DIFFUSION PLANTS: PROPOSED ACCELERATED MEDICAL SCREENING SCHEDULE

Site	No. current workers (CW)	Proposed No. CW screened in next 12 months	Estimated No. former workers (FW) "At risk"	Proposed No. FW screened in next 12 months	Total proposed No. screened in next 12 months
Paducah	1,800	900	7,000+	[1] 1,000	1,900
Portsmouth	2,000	1,000	5,000+	[1] 1,000	2,000
K–25	1,700	850	3,000+	[1] 1,000	1,850
Total	5,500	2,750	15,000+	[1] 3,000	[1] 5,750

[1] We are currently funded to screen 400 of these 1,000 at each site, or 1200 workers in total.

Early Detection of Lung Cancer

Lung cancer is the most important specific cancer risk for workers at the gaseous diffusion plants of the Department of Energy. Occupational exposure to lung carcinogens at the gaseous diffusion plants, including asbestos, uranium, and possibly plutonium and beryllium produce excess risk of lung cancer. If early detection of lung cancer is achievable as a result of medical screening, its implementation should be accorded the highest priority among gaseous diffusion plant workers, especially for those at the highest risk of lung cancer. We do not currently offer such screening in the Worker Health Protection Program.

An effective and feasible method for the early detection of lung cancer now exists. The Early Lung Cancer Action Project, undertaken at Cornell University and New York University Medical Schools, decisively and affirmatively answers the question of whether CT scans of the chest can identify small malignant lung nodules at a sufficiently early stage that surgery can successfully remove the cancer with the expectation of cure. Henschke and colleagues published the results of their landmark study, the Early Lung Cancer Action Project, in Lancet on July 10, 1999. Undertaken with NIH support, this study began in the early 1990's. It enrolled 1,000 people, aged 60 or over, who had a tobacco use history and were sufficiently healthy to undergo chest surgery, if required. All participants underwent a chest x-ray and a low-dose rapid chest CT scan. Lung nodules were identified, and the affected participants were subject to a protocol of conventional chest CT scan and, if relevant, diagnostic work-up.

The study results were remarkable. Low dose chest CT scans detected lung cancer in 27 people (2.7 percent), or in 1 of every 37 study participants. By contrast, malignant lung nodules were seen on conventional chest x-ray in only 7 participants (0.7 percent). Thus, low dose CT scans detected nearly 4 times as many lung cancers as did routine chest radiography.

More importantly, low dose CT scanning nearly always detected lung cancers at an early stage that is usually highly curable. Of the 27 CT-detected cancers, 26 (96 percent) were resectable, and 23 (85 percent) were in the initial stage (Stage I) of lung cancer. By contrast, only about one-half, or 4 of the 7 (57 percent) malignant nodules identified by the chest x-ray were Stage I disease. We know that Stage I lung cancer nominally has a 70 percent to 80 percent 5 year survival compared to an overall 5 year survival of 12 percent for all cases of lung cancer combined.

In addition, only one study participant underwent a biopsy that was specifically recommended by the study protocol and had benign disease. Thus, low-dose CT scanning, when followed by a proper work-up, will result in few people needlessly undergoing the pain and expense of biopsy for benign nodules. The authors conclude: "Low-dose CT can greatly improve the likelihood of detection of small non-calcified nodules, and thus of lung cancer at an earlier and potentially more curable stage." A full summary of this pathbreaking study recently published in *Lancet* is provided in Attachment B.

The results of the Early Lung Cancer Action Project, in combination with current knowledge about the biology, radiology, and epidemiology of lung cancer, are sufficiently convincing to justify the inclusion of low-dose chest CT scanning and an associated follow-up protocol in the medical screening program for gaseous diffusion plant workers. The new lung cancer screening protocol should be offered to gaseous diffusion plant workers who are at highest risk for lung cancer as a result of the occupational exposures to asbestos and uranium and possibly plutonium and beryllium.

We propose to offer such an early lung cancer detection program to 2,000 participants in the Worker Health Protection Program at the gaseous diffusion plants of the Department of Energy. This component will be offered to individuals, both current and former workers, who meet pre-determined criteria for lung cancer risk, as constituted by age, duration and likelihood of exposure to occupational lung carcinogens, and history of cigarette smoking. This program component will be integrated into the existing protocol of the Worker Health Protection Program and, thereby, achieve considerable efficiency and costs savings, especially in participant recruitment, baseline testing, follow-up, and overall program administration.

Medical advances typically benefit metropolitan areas of the United States first, since large cities house the leading medical schools and major medical centers. Lung cancer screening will be rapidly established in New York, San Francisco, and Chicago. Later and perhaps slowly, it will diffuse to rural areas, where DOE facilities are typically located. Through integrating the proposed lung cancer screening method into our Worker Health Protection Program, we have the opportunity to reverse this pattern and make Paducah, Portsmouth and Oak Ridge among the first communities in the nation to receive the great benefits of this life-saving screening technique. The United States Congress and the Department of Energy will accrue enormous gratitude from the current and former gaseous diffusion plant workers as a result of literally saving the lives of a significant number of such workers through supporting lung cancer screening and the Worker Health Protection Program.

LACK OF ACCESS TO OCCUPATIONAL HEALTH CARE: A CORE PROBLEM FOR GASEOUS DIFFUSION PLANT WORKERS

The first core problem in occupational health at the gaseous diffusion plants of the Department of Energy problem is the lack of access of former and current DOE workers to objective, expert, independent care in occupational medicine. When any of us develop a heart arrhythmia, a neurologic syndrome, or cancer, we fully expect to see a physician who will bestow upon us his or her candid, specific, expert opinion that is the distillation of many years of specialized training and clinical experience. We further expect that this opinion will be unencumbered by any conflict of interest of the physician, such as a financial interest in a particular medical tool or laboratory, which would influence the opinion of that physician, sometimes to our detriment. These conditions frame a basic standard of care that we have come to expect in our country.

These conditions, however, do not currently exist, and indeed have never existed, for the workers at the three gaseous diffusion plants of the Department of Energy, or probably throughout much of the DOE complex. Such workers have never as a rule had an opportunity for this simple encounter: to have a potentially work-related illness evaluated by a physician who has the knowledge to determine whether the illness is work-related and is free to make that determination without concern about ramifications to the employer. Instead, workers in Paducah, Portsmouth, and Oak Ridge raise their health concerns with their primary care providers who do not ask about or know about occupational hazards. Or their health concerns arise with physicians who are employed by or under the influence of DOE contractors and thereby have dual loyalties. It is little wonder, therefore, that workers, who are very proud of the service that they have performed for the past 5 decades, nonetheless feel that they have been treated unfairly with reference to occupational illness.

Two immediate consequences result from this failure to provide a basic standard of occupational health care. First, occupational illness is not properly diagnosed and treated. This harms the individual. It also harms co-workers and future workers, because it prevents the return of vital information to the workplace, information that could be used to prevent other workers from becoming ill.

The second consequence is that workers and their families will form their own opinions about whether the workplace is the source of their ills. In the absence of external expert knowledge, workers will use their own expertise to decide about work-relatedness of their problems. Often they will be correct. Indeed, the history of occupational medicine is replete with examples of occupational diseases first identified by workers and later confirmed by physicians. Sometimes, however, workers will not be correct in attributing their symptoms to the workplace. The result of this error is that the DOE facility may be falsely targeted as the source of a spectrum of diverse and quite unrelated illnesses. We cannot blame people who make this judgment: they do so in a vacuum. The underlying problem is the structural lack of a system that can authoritatively and credibly confirm or refute workers' suspicions about workplace exposures as the source of their ill health.

LACK OF ACCURATE EXPOSURE CHARACTERIZATION: A CORE PROBLEM FOR GASEOUS DIFFUSION PLANT WORKERS

Let me turn to a second core problem in occupational health at the gaseous diffusion plants: the lack of proper, accurate information about exposures that have occurred at the gaseous diffusion plants over the past four or five decades. Ultimately, in occupational medicine, we are called upon to make a judgement about whether a health problem of a particular individual is work-related. The equation that rules this decision is quite simple. On the one side is information about the exposure or workplace factor. On the other side of the equation is the delineation of the illness. The latter is usually straightforward given the armamentarium of medical tools that we now have to conduct medical investigations.

The weak link in this equation is often the level and quality of knowledge about the workplace exposures. Chronic occupational illness today results from exposures that occurred in the past. We are therefore subject to whatever actions people who were responsible for the workplace did or did not take to measure those exposures. In 1996–1997, as part of the Worker Health Protection Program, we conducted a one year needs assessment of workplace exposures and the rationale for medical screening at the gaseous diffusion plants (Executive Summary in Attachment C). We concluded, as have others, that workplace exposures have been poorly documented in general at the gaseous diffusion plants, either through failure to measure properly, or through failure to document measurements in a manner that can be properly interpreted. This applies to radiation measurements, but even more so to assessment of hazardous chemical agents such as asbestos, silica, and beryllium.

One important consequence of this failure is that it makes the decision-making about causality between workplace exposures and health problems that occur many years later difficult and complex. When a gaseous diffusion plant worker, or more likely, retiree, develops lung cancer, the likelihood that his prior occupational exposures to asbestos or silica contributed to the development of the lung cancer depends very much on the intensity, duration, and timing of his exposures to asbestos and silica. If information on those exposures do not exist, the amount of judgement that must be used to decide on work-relatedness of that lung cancer increases. And, so too does room for disagreement in formulating that judgement.

A cynical means to "eliminate" occupational disease now becomes apparent. First, on a prospective basis, fail to document exposures in a thorough, reliable, and interpretable manner. Second, overlook communicating meaningful information about those exposures to workers. Finally, decades later, when chronic occupational diseases of long latency appear, claim retrospectively that insufficient data on exposure preclude proper assessment of the causal role of such exposures in the development of the extant illnesses. Note that the premature deaths and diseases suffered by workers do not disappear under such a scheme. But the occupational attribution vanishes.

Let me provide an example relevant to the "discovery" of plutonium, neptunium, and other transuranics at the Paducah gaseous diffusion plants. A memorandum from 1960 has just now been discovered, entitled "Neptunium [237] Contamination Problem, Paducah, Kentucky, February 4, 1960." (Attachment D) It was written by Dr. C. L. Dunham, a physician who directed the Division of Biology and Medicine of the Atomic Energy Commission (AEC), the predecessor to DOE, and a physician colleague from the same Division. Dr. Dunham was therefore the chief physician of the AEC and presumably took the same Hippocratic Oath that every physician takes upon entering the profession. In this memo, they discuss in some detail how neptunium arrives in Paducah, how it deposits on the inner barrier tubes that are the central component of the gaseous diffusion process, and how workers are exposed to the neptunium. They then refer to urine neptunium levels taken in some workers. These physicians further specify that up to 300 Paducah workers should be tested but that, referring to management personnel "they hesitate to proceed to intensive studies because of the union's use of this as an excuse for hazard pay (p. 3)" Dr. Dunham and colleague further argue in favor of the need to obtain post mortem tissue samples, but state that this was difficult due to "unfavorable public relations." Dr. Dunham and colleague conclude: "Thus, it appears that Paducah has a neptunium problem but we don't have the data to tell them how serious it is." There is a striking absence of any formulation of a plan of how to collect those data and how to reduce neptunium exposure at Paducah.

And now, forty years later, we are asked to judge how significant that exposure might have been, who was the population at risk, and whether a retiree's cancer was caused by that unquantified and, presumably, uninvestigated exposure to neptunium, plutonium, and other materials. And at the end of the current spate of urgent investigations, news reports and hearings, there will be some who will conclude ruefully that "we simply do not have the data to tell them how serious it is" and will thereby be paralyzed by this ignorance. I cannot think of a better way to make occupational disease "disappear."

CONCLUSION

Clearly, our present obligations to workers who built and maintained our nuclear weapons stockpile requires that we move beyond paralysis. Towards this end, we have developed a concrete plan to enhance the Worker Health Protection Program. The presence of the Worker Health Protection Program already in place provides an outstanding opportunity for Congress and the Department of Energy to respond immediately to the enhanced need of its gaseous diffusion plant workers for appropriate and timely medical screening for work-related disease. For an additional $5.8 million dollars in the next year, the scope and coverage of the medical testing and education program can be significantly expanded in a well-targeted and clearly justified manner. We will provide comprehensive screening for 5,750 current and former gaseous diffusion plant workers. We will bring the most important advance in cancer screening since the advent of mammography. And this will be accomplished at a fraction of the estimated $1 billion cost that it will take to clean-up the environment at the Paducah site alone.

In conclusion, our program expansion will allow Congress and the Department of Energy to address the concrete and heightened concerns of former and current gaseous diffusion plant workers. Moreover, and most importantly, the advent of a radiographic screening technique for lung cancer will allow Congress and the Department of Energy, through an enhanced Worker Health Protection Program, to save lives.

STATEMENT OF RICHARD CRANSON BIRD, JR., M.D., BETH ISRAEL DEACONESS MEDICAL CENTER, BOSTON, MASSACHUSETTS

Senator MCCONNELL. Thank you, Dr. Markowitz.

I am also going to take Dr. Bird before doing some questions. Dr. Bird is a doctor—Dr. Richard Bird, Jr., is a doctor of internal medicine on the staff at Beth Israel Deaconess Hospital in Boston. He also works with the JSI Center for Environmental Health Studies in Boston. Dr. Bird is one of two doctors hired by the Department of Energy to investigate claims of more than 50 workers at the Oak Ridge, TN, plant who were reporting unexplained illnesses.

He is here today to discuss his findings and to comment on the proposals to expand worker health testing programs. Thank you,

Dr. Bird, and see if you can come close to 5 minutes some opportunity for questions.

Dr. BIRD. I will do my best, and I will skip my own background information.

EXPOSURE AND EVALUATION OF K–25 WORKERS

It is my understanding that I have been invited here today to present a more general summary of some of the findings that have been formulated to date in evaluating workers from the Oak Ridge, TN, K–25 facility and to identify those areas of work which will be forthcoming and of potential importance to this committee.

I thank you for the opportunity to present this information today and will begin with some background. In the latter half of 1996, I was asked by representatives of Lockheed Martin Energy Systems to participate in an evaluation of workers at the K–25 facility in Oak Ridge. I was invited to collaborate with Dr. James Lockey, Director of Occupational Medicine at the University of Cincinnati, who has been a member of the Fernald Workers Settlement Fund Expert Panel. In addition to his extensive experience in the field of occupational and environmental medicine, Dr. Lockey specializes in pulmonary and internal medicine.

Several workers at the K–25 facility had developed symptoms and conditions that they were concerned may have been related to exposures at work. I had answered a few phone inquiries made to JSI in the summer of 1996 from workers representatives who had asked specific questions about testing these workers. My responses led to a formal request to participate in this overall evaluation process.

Some of the workers at the K–25 facility had been evaluated by on-site medical department personnel because of health concerns. One provider had performed measurement of urinary thiocyanate as a marker of cyanide exposure using what we later confirmed was an outdated and unreliable method. The results varied widely and included several values that were reported as elevated.

The National Institute of Occupational Safety and Health, NIOSH, was asked to respond to the possibility of cyanide exposure and took several air samples in the areas where workers were found to have elevated urinary thiocyanate measurements. NIOSH did not identify corresponding elevations in airborne cyanide. However, the question remained whether other factors had contributed to various symptoms and conditions experienced by several workers.

Dr. Lockey and I began a series of meetings to initiate an individual evaluation process involving over 50 workers who had asked to participate, with the goal of attempting to determine whether workplace factors had contributed to the symptoms and illnesses of each individual case. Dr. Freeman was brought into the process by Dr. Lockey to assist with the extensive work involved. He is currently on staff in Occupational Medicine in Cincinnati as well and has some background in neurology.

At the outset, Dr. Lockey and I asked to be allowed to arrange for independent industrial hygiene measurements or studies at the K–25 facility in the event that during the process we thought this may be of use to us in specific areas. We learned very quickly that the complexity of this site has been so vast that it has not been possible for us to independently recreate the industrial history of each area of potential exposure concern.

The K–25 facility has operated as a gaseous diffusion plant for uranium purification since World War Two until the mid-1980's. This has included an extensive infrastructure of support and research operations, some of which interface with the Y–12 production facility and some which involve characterizing, storing, shipping, reclaiming, and incinerating hazardous materials from various sources. Some industrial hazard management and research activities continue today.

Representatives from Lockheed Martin Energy Systems and now Bechtel-Jacobs have been very helpful in attempting to answer questions we have had, and we have pursued independent industrial hygiene studies of specific areas, however with a limited scope. We have also benefited from risk characterization summaries prepared for the former worker surveillance program. Most importantly, however, we have benefited from histories provided by individual workers detailing their work experience.

Many of the workers in our group were involved—in our group we were evaluating, were involved characteristically in industrial activities that brought them throughout the facility. Several began working at K–25 in the 1970's. Some worked outside for shorter periods of time, most often at the Y–12 facility.

A smaller portion of the workers in our group were administrative, technical, managerial, or service employees who were not involved directly in industrial activities. This has raised questions about whether they might also have been at risk for potentially significant exposures. Some of these workers were located in former industrial areas or performed jobs nearby ongoing industrial or hazardous storage areas, while others were only located in non-industrial buildings. Targeted questions have been pursued regarding the possibility of hazardous materials in non-industrial areas, some of which have been addressed and others are still under review.

By June of this past summer, a procedural protocol was completed to allow us greater access to plant-wide areas, which has expanded our understanding of materials and handling operations. Despite the complexity of site assessment issues, we have worked predominantly on individual medical evaluations, which provides the framework for our determination of potential work-related illness.

The principal medical process that we have pursued has involved detailed consideration of each individual case, including an extensive review of available past and ongoing medical records, medical histories, including social, family, and occupational histories, physical examinations, and referrals for evaluations and diagnostic studies, including both markers of effect and markers of exposure when available. This has been particularly difficult regarding past exposures.

This process requires an intimate, personal and confidential relationship with individual workers and requires attempting to interface with treating physicians either already involved in an individual's care or integrated through our referral recommendations. Further complexities include interim disability determination, insurance management, and financial constraints for travel to referrals, much of which has been improved on with the help of Lockheed Martin Energy Systems, Bechtel-Jacobs, and I believe Department of Energy personnel.

BASIS OF DETERMINATION

The principal basis for our determinations has been and will be based on our detailed review of individual symptoms and conditions, with particular importance placed on chronology in relation to work histories and ability or not to establish corresponding diagnoses of patterns of illness. Of equal importance are pre-existing medical histories, predisposing factors, and the possibility of changes in disease patterns in relation to work.

We are attempting to determine within a reasonable degree of medical probability and certainty that an individual's symptoms and corresponding conditions are or are not likely to have been significantly impacted by or due to exposures from working at the Oak Ridge K–25 or other facilities or whether this is unknown at this time.

In some cases we are not able to determine whether illnesses have been significantly impacted by work factors for several reasons, including a lack of adequate medical knowledge within the scientific literature on specific occupational exposures present at these facilities.

At the time of our conclusions, we will attempt to identify either areas that are in need of further study from a basic science and clinical epidemiology perspective.

Most of the workers have presented with several symptoms and conditions, which we have reviewed and summarized in the form of illness categories in interim and, more recently, update reports. These have included referral recommendations linking individuals to treating medical providers. In some cases this has led to important treatment interventions regardless of work-related considerations.

To date, we have identified several individuals who are likely to have developed respiratory illness impacted by work place exposures. Some of the diagnoses have included chronic rhinitis and sinusitis, chronic bronchitis and occupational asthma. We have identified that approximately 10 percent of the workers of our group have developed sensitization to beryllium, some of whom worked mostly in non-industrial areas. A few of these individuals have also developed actual beryllium-related lung disease. We have recommended further characterization of the K–25 facility to attempt to identify and remediate areas with beryllium contamination.

We have also identified significantly elevated levels of airborne molds in a major hazardous waste storage building and recommended that respirator protection be utilized.

Senator MCCONNELL. Dr. Bird, I am sorry. Let me just say, with all due respect to all the witnesses, we are never going to finish if we cannot do 5-minute summaries of the statements and have an opportunity for a few questions. So if you do not mind, what I would like to do is put your entire statement in the record. If there is a way for you to finish it in a minute or so, that would be great.

Dr. BIRD. I will be glad to do that. Thank you.

We have identified some workers with neurologic illnesses that are likely to have been impacted by work place exposures, including peripheral neuropathies, brain function impacts, and psychological impacts. We have also identified several workers who have developed secondary psychological conditions impacted by concerns over work place exposures, concerns about deteriorating health, and difficulties associated with working in potentially hazardous settings.

I will skip to really what we intend to present to the public in the future relative to these matters. In concluding this presentation today, I would like to outline those areas of importance which can be derived from this extensive clinical undertaking and which we hope to address in more detail when we present our conclusions publicly.

The Oak Ridge facility operated during several decades for vital purposes. While this facility has greatly transitioned during more recent years, the workers here, both past and present, are in need of all that can be done, all that can be offered to assist with the potential impacts of activities at this site.

It is appropriate and commendable that the U.S. Department of Energy, Lockheed Martin, and Bechtel-Jacobs have pursued an in-depth and independent clinical response here for purposes of helping several workers with various unexplained illnesses and symptoms.

Individual care requires individual medical evaluations. It is often difficult, however, for any patient with a potential occupational illness to identify providers who have the time or background to consider work place factors. In those industrial settings which are extremely complex and pose uniquely concerning hazards, it is especially important to consider the use of clinical evaluators knowledgeable about the type of exposures that can exist in these types of facilities to develop an approach that can be applied to other workers or community members.

We anticipate presenting a more detailed summary of work-related illnesses identified at the Oak Ridge facility. We will elaborate on more difficult areas that raise further questions, including limitations of medical knowledge for diagnostic and clinical purposes, and will make suggestions about additional studies that may be helpful.

We will summarize any specific exposure concerns pertaining to this site and make recommendations for further assessments of areas that we identify as important. We will be available to collaborate with Dr. Markowitz and others to help determine whether general surveillance should be expanded based on our findings, and we will consider recommendations for conveying our findings to regional medical providers, including locations that have the capability for in-depth evaluations of other workers and for those involved in providing ongoing care.

Thank you very much.

Senator MCCONNELL. Let me just say, in fairness to all the witnesses, we have a lot of witnesses and we have a lot of questions. So what we are going to do after this panel is 5 minutes means 5 minutes; the hammer is coming down. It does not mean anything you have to say is going to be lost, because we are going to have an extensive record here.

If I am unable to ask all the questions, they will be submitted to you in writing, because we want a complete record, and we will ask you to return those answers within a couple of weeks.

CHANGE IN DOE PROGRAM OFFICE STRUCTURE

Mr. Fuller, you indicated in your testimony that you share my view that the Paducah and Portsmouth program office should be moved out of Oak Ridge. Could you explain how the workers might benefit from this proposal?

Mr. FULLER. I think, Mr. Chairman, I think there would be a benefit in that effort, cleanup efforts at Paducah, we could have a management team on site at Paducah that could focus themselves on the particular jobs that are things that need to be accomplished at Paducah. I think that with that we could be a lot more effective. I think we could save money, and I think we could get the attention we need. We could direct the proper job at the proper time with the proper people, and I believe we could do it more safely and with better oversight.

So those are part of the reasons that I would support that.

DOE COMPENSATION PROGRAM AT PADUCAH

Senator McCONNELL. Your testimony cites a number of examples of Federal compensation programs that have been established to help workers who served their country in connection with the nuclear weapons program. Prior to the lawsuit which generated all the attention at Paducah, what level of compensation program was being offered to the workers at the gaseous diffusion plants, and did you ask the Department to establish a plan for your members?

Mr. FULLER. We have not—we did not. The only recourse that we have ever had has been through the State, State recourse, and of course that is almost impossible to recover anything through the State, because of the burden of proof problem, to have to show causation. We have naturally not been able to do that.

And no, we had no request or special request prior to this for that.

SITE REGULATORY RESPONSIBILITIES

Senator McCONNELL. Have you noticed a change in USEC operations as a result of the Nuclear Regulatory Commission taking over regulatory responsibilities from DOE?

Mr. FULLER. Yes, Mr. Chairman, as a matter of fact I have. I do not know that I have any data to back this up, but I can certainly tell you how it appears to me, working in the site and being there every day. In my judgment there is considerable more stricture, more procedure, more emphasis on those things, on procedure and safety concerns.

Just, if I might say, it just seems to be a tighter ship safety-wise and in regard to how workers do their jobs and how the plant is run.

Senator MCCONNELL. Would you support bringing in an independent regulator to ensure that DOE's worker health and safety standards are not compromised?

Mr. FULLER. Certainly, yes. The last thing the union wants is to see any compromise on worker health and safety, and we would support anything that would ensure that.

Senator MCCONNELL. You noted in your testimony that the existing management and integrator contract has compromised worker safety. This position is supported by the phase one investigation. You noted that the problem could be solved by hiring experienced workers.

The question is what have DOE and Bechtel-Jacobs done to help transition the experienced workers into these cleanup jobs?

Mr. FULLER. Well, Mr. Chairman, that has been a point of ongoing problems. We have been endeavoring over the past few months to work with Bechtel-Jacobs to come to some agreement that would transition incumbent workers into the cleanup efforts. To this point, we have only been able to get Bechtel-Jacobs to discuss 28 jobs with us for transition. That is with the full knowledge that there is a huge amount of work that will be done out there in the near future, one hopes, and we would want to be able to transition the incumbent work force into those jobs, for what we think are obvious reasons.

We have a huge amount of institutional knowledge in that incumbent work force, people who are familiar with the site, familiar with the problems, and could be of great value if we can transition those folks into the jobs.

I will say that we have had a hard time getting that done. One of the problems, of course, is the timing of the transition. There is a good chance there may be layoffs at that site, and that is this summer. If we could coordinate the transition of workers to the cleanup side of the house in the same time frame that people may be losing their jobs due to layoffs with USEC, we could probably accomplish that transition more smoothly.

Senator MCCONNELL. Do you feel that the Nuclear Regulatory Commission, which regulates USEC activities, is more responsive to worker concerns or is the Department of Energy, which regulates its own activities, more responsive?

Mr. FULLER. We have found NRC to resist direct interaction with the unions to a large degree. That is, they do not officially include us on their mailings or copy us on information. We are not routinely a part of their outbriefings and so forth.

So they do not have a place for us in their scheme of things. They deal with management and they expect us to deal with management. They do not deal directly with us. That is a bit of a problem. I would like to see a situation where they would include the union and the worker representatives, see them have an interaction with the NRC in the future, if there is some way we could do that. We miss that.

LUNG CANCER SCREENING PROGRAM

Senator MCCONNELL. Mr. Fuller, I want to reiterate my strong support for the efforts of Dr. Markowitz to help workers identify and receive treatment for the illnesses they might have contracted while working at the plants. I also strongly support his proposal to expand the testing program to include current workers and to begin the early detection lung screening program.

I was pleased that the Energy and Water Appropriations Conference Report for fiscal year 2000 included language that I added requesting that the Department expand the program to provide funding to begin the lung screening effort, which as you indicated, is so valuable in detecting early stage lung cancer.

Dr. Markowitz, you indicated in your testimony that in order to test the 15,000 former workers at the three sites based on the Administration's requested funding level it would take 12 years. If the goal of the program is to help workers identify illnesses as early as possible, why do you suppose the funding level is so low?

Dr. MARKOWITZ. I think they probably arrived at an overall figure for the program nationwide and divided it by the number of sites that they wanted to cover. I do not think that the budgetary figure actually has any relationship to need. If it were, there would have been a different process of matching up what we know about exposures that people have had, what they are at risk for, and what that means in terms of budget. In other words, one would have designed a medical program and then looked at budgetary allocations that matched that program. That was not the process as far as I know.

Senator MCCONNELL. Do you know who decided to exclude current workers and what the basis for that decision was?

Dr. MARKOWITZ. I know from the beginning that the emphasis was on former workers. I am not certain whether current workers are absolutely excluded by DOE, but I know all along that the emphasis was on former workers, in part because they may be at higher risk. They worked earlier years at the plant and they have had a longer time to develop occupational illnesses.

But clearly, many of the current workers have also had a long period of time and have also suffered exposure conditions that would lead to disease.

Senator MCCONNELL. What would it cost to expand the testing program to 5,750 workers a year? Do you have a budget estimate for that?

Dr. MARKOWITZ. Yes. It would cost $5.8 million. We currently receive $1 million. For an additional $5.8 million, we will go from testing 1,200 workers per year at the 3 gaseous diffusion plant sites to 5,750 workers per year. In addition, we would include CT scanning for the early detection of lung cancer.

Senator MCCONNELL. Would that include purchasing new equipment, the figure you just gave?

Dr. MARKOWITZ. Yes. The $5.8 million includes purchase of a CT scanner and a mobile unit to transport the CT scanner among the various sites. Those equipment costs are about $1.4 million, so actually the cost in subsequent years would be less of actually operating the program because we would have the equipment at that point.

Senator MCCONNELL. And at that level you could complete the whole thing in how many years?

Dr. MARKOWITZ. We could offer to current workers, we could offer the program to all the current workers, within 2 years. For the former workers, it would take probably closer to 4 or 5 years.

Senator MCCONNELL. After listening to Dr. Bird's testimony, how do your findings compare to the research compiled by him?

Dr. MARKOWITZ. They really are sort of different areas. Dr. Bird and Dr. Lockey focused on several dozen individuals who were ill, claimed illness from workplace exposures, and a variety of illnesses. Heretofore, we focused on diseases for which section 3162 mandated screening, that is to say conditions that we could identify early for whom we could do some good, and in complying with that we have focused on chronic lung disease, kidney and lung disease and hearing loss.

Now, I will admit that does not cover the whole gamut of occupational illness. We could not do that for that budget. But in addition, all of occupational illness is really not amenable to early intervention and identification. So I think to some extent there is some overlap, but also some mutually exclusive aspects to his work in what we are screening for.

IMPEDIMENTS TO IDENTIFYING RISKS AND SOURCES OF EXPOSURE

Senator MCCONNELL. For both Dr. Markowitz and Dr. Bird, what are the greatest impediments you have come up against in identifying risks and source of exposure to the work force? More specifically, what information could the Department or the contractors provide to you that would assist you in diagnosing workers? You want to lead off, Dr. Bird?

Dr. BIRD. Well, I think that exposure information, as I tried to elaborate on somewhat in my testimony, is a very difficult topic, because the greatest information comes from the work experience of individual workers. There certainly was a very impressive industrial hygiene operation on site and there is a lot of information that is so vast that it is impossible for me to, or Dr. Lockey and our team, to have an ability to fully understand all that is going on here. You are talking about 40 years and 50 years of very detailed monitoring throughout the site.

So the real question is can we identify diseases that are likely to be related to the exposures that we are concerned about here, and that is our task. This is a very different process than what can be done in screening, and I think that we can learn quite a bit from having come in from the outside because of the independent relationship with patients, independent of fears about employers and revealing things that they feel might jeopardize their job security or things of that sort. We can identify personal non-work-related and discuss those illnesses as well in that process.

Senator MCCONNELL. Anything to add on that?

Dr. MARKOWITZ. Well, we have gotten excellent cooperation by all parties, including the contractors, certainly the local unions, Department of Energy, both at the sites and at the central office. I know Assistant Secretary Dr. Michaels is supportive of the program.

Part of the problem with exposure data which is really key is that we do not know entirely what is there. We conducted a 1-year needs assessment for the three sites about 2 years ago and we profiled what we believe to be available exposure data, characterizing what people have been exposed to over the previous years. We knew those data were inadequate, so we did risk mapping of our own, taking groups of workers and mining their collective memory to look at what kind of exposures have occurred over the past several decades.

Now we find out in August of this year that there was contamination with plutonium and neptunium that we did not know about. So in some respects it is a question of there being data that exist that we simply do not know about, so we do not know to ask for that.

We would like a complete cataloguing of those data that exist and having access to them.

EXPANSION OF HEALTH RESEARCH EFFORTS

Senator MCCONNELL. Dr. Markowitz, you have been working with the Kentucky School of Public Health, which is a joint effort between the University of Louisville and the University of Kentucky. Could you update us on the opportunities to expand the health research you are currently considering and how might this benefit the work force and their families and the community?

Dr. MARKOWITZ. We have had communications, several in fact, with physicians and others from the University of Kentucky in Lexington, the medical school, and also the University of Louisville, the medical school. These have been excellent discussions. We would like very much to include them in our program. They want to be one of the clinical facilities doing the screening at Paducah and if we get the expanded funding we will be glad to include them in that.

We think they can play a central role in creating what does not exist right now, which is a diagnostic and treatment center for patients, for former and current DOE workers, in Paducah. By the way, I think Portsmouth and Oak Ridge need access to such a center as well. I am speaking about a regional center of excellence in which physicians would be able to provide honest, independent, expert opinion about the diagnosis and treatment of occupational disease. Workers in Paducah should have access to that.

We are providing screening. Ours is a one-time screening to identify people who need further diagnosis and treatment, and they need that kind of resource, and I think that the medical schools in the State should absolutely be involved with that.

There may be some research opportunities to collaborate with them and we are certainly receptive to that, as long as it is clearly in the welfare of the workers at the Paducah site.

Senator MCCONNELL. Finally, to Dr. Bird and Dr. Markowitz both, have you reviewed the 1983 autopsy report on Joe Harding? What are your conclusions and have you seen other workers exhibiting similar conditions?

Dr. BIRD. I have not reviewed that, no.

Dr. MARKOWITZ. I have not reviewed that in depth yet, either, so I really cannot comment on that.

Senator MCCONNELL. All right. Well, I want to thank all three of you for being here this morning. We appreciate it very much.

STATEMENT OF DAVID MICHAELS, PH.D., ASSISTANT SECRETARY FOR ENVIRONMENT, SAFETY AND HEALTH, DEPARTMENT OF ENERGY

ACCOMPANIED BY:

DR. DAVID STADLER, DOE'S ACTING DEPUTY ASSISTANT SECRETARY FOR OVERSIGHT

BILL ECKROADE

Senator MCCONNELL. We would like now to call Dr. David Michaels, the Assistant Secretary of the Department of Energy's [DOE] Environmental Health and Safety Program, who will present the findings of the DOE phase one study. Dr. Michaels has been in his current position for about a year and is also the DOE official overseeing the investigation at the Paducah plant. Before joining DOE he was a professor of community health at City University in New York. He is also an epidemiologist, with more than 20 years of experience in public health, particularly occupational and environmental health associated with the impact of industrial operations.

We are pleased to have you here, and let me say again, at the risk of appearing heavy-handed, 5 minutes means 5 minutes, and not a single pearl in your statement will be lost. It will all be part of the record, and that will give us an opportunity to have some questions.

Go right ahead, Dr. Michaels.

Dr. MICHAELS. Thank you, Senator McConnell. I greatly appreciate the opportunity to discuss the results of the first phase of DOE's independent investigations into allegations of environment, safety, and health concerns at the Paducah gaseous diffusion plant.

As you know, Secretary Bill Richardson committed to conduct a complete and independent investigation of these allegations, and this report represents the first installment on that commitment. Detailed results from the investigation, released last week, are provided in my written statement and in the report itself.

With me here today are Dr. David Stadler, DOE's Acting Deputy Assistant Secretary for Oversight, and Mr. Bill Eckroade, who led the environmental section of the investigation.

This investigation was conducted by senior investigators and technical experts from my staff. We will be planning to do similar investigations at the gaseous diffusion plants in Tennessee and Ohio shortly.

I am just going to go through this very fast. We divided this investigation into two phases, so we can give you the first results pretty quickly. The second phase is under way, focused on historical environment, safety, and health performance, that is before 1990, and we hope to have that investigation completed by January 2000.

At the outset let me say that the investigation team found no immediate threat to Paducah workers or to the public that would require the plant to be closed down. Cleanup is being conducted in accordance with an agreement among DOE, EPA, and the Commonwealth of Kentucky and the site is currently in compliance with that agreement.

The team noted that since the early 1990's steps have been taken to protect the public and mitigate the impact of radiological and chemical contamination, such as hooking up homes with public water, so the current risk from this contamination is not high. Actual radiation exposures to workers have been low and injury and illness rates at the Paducah site are lower than at many other DOE sites.

RESULTS OF PHASE ONE STUDY

At the same time, however, the team identified a number of problems that, viewed together, are cause for concern, management attention, and corrective action. The team concluded that, while the site is in compliance today, its inability to meet upcoming major cleanup milestones under that agreement is threatened.

Work to date has been limited largely to characterizing contamination, operating and maintaining the site, meeting regulatory requirements, and controlling the spread of contamination. Most contamination sources identified in 1991 still remain. Ground water contamination plumes now extend more than two miles off site and continue to grow, and the site has not adequately characterized these plumes.

The team concluded that significant steps to improve protection of the public and the environment are needed to avoid the possibility of health risks in the future. Management needs to emphasize actual remediation that addresses continuing sources of contamination, to limit degradation of contaminated buildings, and to control the continued spread of contamination.

In the area of radiation protection, the team found that since the 1990 tiger team report, progress has been made. While the investigation team identified similar deficiencies today, the magnitude of these problems is less. Records indicate the external doses to employees from the types of radiation present at Paducah are very low and there have been no significant up intakes of radioactive material.

Radiological protection problems found today are typical of a site that has had to cope with legacy hazards for many years and which is no longer an operating facility. There has been increasing reliance on worker knowledge rather than a disciplined and rigorous application of controls. These weaknesses are worsened by the lack of effective DOE or Bechtel-Jacobs oversight of radiation work practices.

Criticality safety deficiencies in storage areas, 148 areas where large amounts of legacy materials are stored across the site, pose an unnecessary hazard to workers in the surrounding areas. These materials have not been characterized fully and 11 of them contain potential fissile material deposits. As a result, the risk of inadvertent criticality, while remote, is not known.

Finally, the team reviewed the quality of oversight ES and H activities at Paducah by the Department and its contractors. The current effectiveness of line management oversight of environment, safety and health and assurance of compliance with DOE requirements is a matter of concern.

In response to this report, line management has developed interim corrective actions, such as providing additional radiation protection training and dosimetry for subcontractors, increased posting of contaminated areas, and precautions to further limit the potential for criticality accidents. Further, DOE offices at both head-quarters and the field are developing detailed corrective action plans, to be submitted within 30 days.

I want to emphasize that Secretary Richardson takes the concerns that have been raised seriously and is committed to investigate and resolve them. We have much work in the months ahead as we complete the second phase of this investigation.

Mr. Chairman, my testimony also describes the status of several activities being managed by my office that were initiated by Secretary Richardson in response to the concerns in Paducah. These include the study of the flow of recycled materials throughout the DOE complex, a worker exposure assessment project to help inform Paducah workers and workers at Portsmouth and Oak Ridge about their exposures, and the expanded program for medical monitoring for both current and former workers as described by Dr. Markowitz.

PREPARED STATEMENT

As you know, despite your best efforts, for which we are deeply grateful, funds for these activities requested by the Department in the budget amendment earlier this year were not provided. Indeed, the budget for my office was reduced significantly. As a result, we are having to defer progress on a number of these activities, especially when they involve contract support, until we are able to identify a source of funds. These remain very high priorities for the Secretary. He is committed to work with you and the committee to find sufficient funds.

Thank you for this opportunity to testify.

[The statement follows:]

PREPARED STATEMENT DR. DAVID MICHAELS

INTRODUCTION

Thank you, Mr. Chairman. I appreciate the opportunity to present the results of the first phase of the independent investigation into allegations of environment, safety and health problems at the Paducah Gaseous Diffusion Plant (PGDP) in Paducah, Kentucky. As you know, Secretary Richardson committed to conduct a complete and independent investigation to determine if any of these allegations were true. He further committed to determine if workers were made ill because of inadequate worker protections and that if they were, to seek to provide them with fair compensation.

DOE is currently responsible for environmental cleanup of waste generated prior to 1993 when the facilities were leased to the United States Enrichment Corporation (USEC), and for the management of the inventory of depleted uranium hexafluoride (UF6) stored at PGDP. This work involves approximately 94 employees of Bechtel Jacobs, the DOE current contractor for cleanup at the Paducah site, a transient subcontractor work force of up to 300 workers, and a small number of workers for USEC that support site cleanup or management of the inventory of depleted uranium hexafluoride. Uranium enrichment activities were transferred to USEC in July 1993 in accordance with the Energy Policy Act of 1992. Uranium hexafluoride and worker safety issues are covered under the authority of the Atomic Energy Act with oversight by DOE. USEC is subject to NRC regulation.

Because PGDP is a designated Superfund site, cleanup is being conducted in accordance with a Federal Facilities Agreement (FFA) among DOE, the Environmental Protection Agency, and the Commonwealth of Kentucky. This agreement establishes milestones and a schedule for meeting them. DOE and its contractors have managed the PGDP under the FFA since the mid-1980s, and the Paducah site is currently in compliance. The investigation found no immediate threat that would require cessation of all plant activities. The current risk to the public is not high, radiation exposures to employees have been low, and injury and illness rates at the Paducah site are lower than at many other DOE sites.

GENESIS OF THE INVESTIGATION

In May 1999, the Department became aware that a qui tam case would be filed under the False Claims Act in U.S. District Court. This suit alleges fraud on the part of contractors at the Paducah Gaseous Diffusion Plant, based on current and past environment, safety and health violations. Once the case was filed, it was placed under a court seal that prohibited DOE from acknowledging or discussing the case with any party outside the federal government. While the allegations could not be discussed, the Secretary felt it important to ensure that there were no imminent threats to the environment, public health or safety and sent a technical team of radiation safety professionals, health physicists and environmental engineers to conduct an on-site review of the areas currently under DOE's control. No public dialogue could be initiated at that time because of the restrictions imposed by the court seal. In August, many of the allegations became widely reported in the national media and Secretary Richardson called for a comprehensive response to the public allegations. The court seal was subsequently lifted allowing the Department to publicly discuss its responses to the allegations.

Many of the concerns regarding worker safety and health stem from the presence of plutonium and other radioactive materials at PGDP and the question of whether workers were adequately informed or prepared to handle such materials. These materials resulted from the recycling of uranium from weapons production plants to the gaseous diffusion plants during the 1950s, 1960s, and 1970s. Concerns are focused on the transuranic elements and fission products that were and are present in this recycled uranium. It is estimated that approximately 100,000 tons of recycled uranium were processed at the Paducah plant.

Environmental concerns alleged in the suit include both on-site and off-site contamination from legacy radioactive or hazardous materials, and the potential for harm to workers or public health and safety. Allegations include:
  —possible improper disposal of hazardous or radioactive materials both on- and off-site in publicly accessible areas;
  —apparent inappropriate release of materials that were radioactively contaminated, release of contamination into site streams and drainage ditches, claims of inadequate control and posting of offsite contaminated areas, and
  —suspected exceedences of radiological air emission standards.

CONDUCT OF INDEPENDENT INVESTIGATION

The comprehensive investigation into environment, safety and health (ES&H) concerns at PGDP is being conducted by a senior team of investigators and technical experts from my staff. The PGDP investigation will be followed by similar investigations at the other Gaseous Diffusion Plants in Oak Ridge, Tennessee and Portsmouth, Ohio. The PGDP investigation was divided into two phases so that we would be able to provide a timely assessment of the current state of environmental protection, and worker and public health and safety. The purpose of the first phase was to determine whether current work practices for those areas of the site that are the responsibility of DOE are sufficient to protect workers, the public, and the environment. The second phase is currently underway, and is evaluating environment, safety and health performance and concerns with historical plant operations from its inception through 1990. We expect that investigation to be complete in January, 2000.

The scope of the first-phase investigation included: facilities and properties under DOE jurisdiction; ES&H issues associated with these facilities and properties from 1990 to present, including interactions between DOE and stakeholders; and ES&H issues associated with uranium enrichment facilities from 1990 to 1997—the point when NRC assumed regulatory oversight of the gaseous diffusion processes, facilities, and personnel. The DOE-controlled operations that were examined included: landlord infrastructure; legacy and newly generated waste treatment, storage, and disposal; site remediation; uranium hexafluoride cylinder storage; facility decontamination and decommissioning; and TCE and polychlorinated biphenyl (PCB) collection, treatment, and cleanup. The investigation did not examine areas leased by the United States Enrichment Corporation (USEC) that are under Nuclear Regulatory Commission (NRC) jurisdiction.

The investigation team gathered information in a number of ways, including: interviewing personnel; observing work activities and performing walkdowns of facilities, work areas, and the site grounds; conducting groundwater, surface water, sediment, and soils sampling; conducting radiological surveys; and reviewing documents. More than 100 interviews were conducted with DOE Headquarters, Oak

Ridge Operations and Paducah Site Office personnel; Bechtel Jacobs and subcontractor managers, supervisors, and workers; selected USEC personnel; and stakeholders. The team also reviewed hundreds of documents including plans, procedures, and assessments that provided perspectives on the assignment of roles and responsibilities, conduct of work activities, and the record of assessment findings.

The Investigation Team collected more than 30 samples from groundwater wells, surface water sources, sediments, soils, and from materials, equipment, and facilities. Samples were collected both inside the security fence as well as on DOE property that is outside the fenceline perimeter. These samples were evaluated for the presence of radioactive and non-radioactive contaminants.

Investigation Results

The team noted that a number of significant environment, safety and health improvements had been achieved since the early 1990s. Since the mid-1980s, steps have been taken to protect the public and mitigate the impact of radiological and chemical contamination, such as hooking up homes to public water. In the worker safety area, there have been enhancements to the radiation protection program, radiation exposures to employees have been low, and injury and illness rates at the Paducah site are lower than at many other DOE sites.

At the same time, the team identified a number of weaknesses in each of the areas reviewed. While the team found no immediate threat that would require cessation of site activities, it found the cumulative impact of the deficiencies to be a cause for concern and corrective action. The results of these evaluations are presented in three main categories—Public and Environmental Protection, Radiation Protection/Worker Safety and Health, and Line Oversight.

Public and Environmental Protection

Industrial operations at PGDP have produced large quantities of legacy materials that have been disposed of in landfills or burial grounds, released into the environment, or placed in long-term storage. Current DOE operations at PGDP focus primarily on the administration of programs to address these legacy materials and on infrastructure maintenance. The team found that cleanup plans and strategies have been developed in accordance with federal environmental regulations and the site is currently in compliance with the provisions of the Federal Facilities Agreement.

Investigations conducted in 1990 and 1991 reported that the PGDP-contaminated offsite groundwater plumes are some of the largest in the DOE complex. Radiological and chemical contamination has spread from the site boundary into the groundwater and surface sediments, particularly into the Big and Little Bayou Creeks. Contamination continues to migrate from sources into the environment. Numerous locations of radiological and chemical contamination have been discovered on DOE property both on-site (within the plant security fence), on the DOE property outside of the plant security fence, and in "offsite" areas now managed by the Kentucky Fish and Wildlife Service.

The plant has taken effective interim steps since 1990 to protect the environment and public health. Groundwater pump-and-treat efforts have helped to impede some of the highest areas of contamination, and alternate sources of water have been provided to residents with contaminated wells. These steps have slowed the spread of contamination from the site to the surrounding environment and reduced public risk, but contamination sources still exist, and the groundwater plume has continued to spread from the site. In addition, actions have been taken to control waste management activities at the point of generation and in the facilities subject to external regulation.

While the current risk to the public is minimal, the team determined that significant improvements are needed in environmental protection.

Findings:

1. Although the site is in compliance with the FFA, there has been limited progress in remediating and characterizing environmental contamination, low level waste, and stored hazardous materials produced by industrial activities. The meeting of major cleanup milestones under the Federal Facilities Agreement is jeopardized by inadequate funding. Work has been largely limited to characterizing contamination, operating and maintaining the site infrastructure, meeting regulatory requirements, and controlling the spread of contamination. Many of the areas of significant radiological and environmental contamination have been identified during past investigations and are the subject of existing compliance agreement.

—Most of the sources of contamination identified in 1991 still remain. Contaminated materials from burial grounds, old landfills, inactive waste lagoons, or spill sites identified in 1991 have not been removed or treated. Groundwater plumes containing trichlorethelene (TCE) and technetium-99 resulting from these source areas continue to propagate at one foot-per-day and now extend for over two miles.

—Contaminated process buildings, shut down more than 20 years ago with no possible future use, have not been adequately maintained or removed. These buildings still contain hazardous materials and have been allowed to deteriorate; they are subject to animal infestation, broken windows, and leaking roofs, are not included in the current cleanup schedule, and are increasing in risk and cost to decommission.

—A large volume of contaminated waste materials at Drum Mountain and scrap metal that has accumulated since the 1950s is stored outside. These areas continue to contribute contamination to the environment through surface water runoff and dispersion. The Federal Facilities Agreement requires removal of this material from Drum Mountain and beneath it by 2003, but current target funding levels threaten reaching this milestone.

—An equivalent of 31,000 55-gallon drums of low-level waste are stored onsite at Paducah, much in containers that were not designed for long-term storage. Many of the containers stored outside are severely degraded, and some have leaked due to this degradation. Much of this waste has yet to be fully characterized—only 157 cubic meters have been shipped from the site since 1990, and the schedule for completion of disposal has been delayed from fiscal year 2006 to fiscal year 2012.

—The 148 DOE Material Storage Areas (DMSAs) located across the site that contain large amounts of material that has yet to be characterized. These areas are not being managed pursuant to either the CERCLA or the RCRA .

—The nearly 37,000 uranium hexafluoride (UF6) cylinders stored onsite in the open at the Paducah plant constitute a radiological exposure hazard and a potential threat to worker and public health in the event of fire and rupture. The Defense Nuclear Facilities Safety Board Recommendation to upgrade the condition and convert the UF6 to a more stable form has been impacted by the cancellation of painting 1,400 cylinders due to funding constraints. Funds have not yet been appropriated for a UF6 conversion facility.

2. There are continuing weaknesses in the radiation protection management of known environmental contamination areas by both Bechtel Jacobs and DOE. While the areas of most significant radiological contamination have been identified during past investigations, deficiencies in radiological characterization, posting, contamination control, and application of environmental as-low-as-reasonably-achievable principles remain. While these conditions don't present a current health risk, such weaknesses violate sound health physics practices. Some examples include:

—The full extent of radiological contamination on DOE property (both inside and outside the site security fence) has not been characterized. For example, at a recently identified area of contamination adjacent to a landfill, a radiologically-contaminated tar-like substance was discovered and subsequently covered and posted to control access. There is no documented listing or database of radiologically-contaminated areas other than what is included in the Solid Waste Management Unit listings, which are not maintained by the radiological control organization and do not clearly designate contaminants of concern for each Solid Waste Management Unit.

—Areas with levels of contamination that exceed Bechtel Jacobs radiological posting criteria were noted on DOE property at some distances beyond the site security boundary. Under the Bechtel Jacobs health physics procedures, these areas should have been posted as soil contamination areas with appropriate measures taken to prevent inadvertent entry. Some of these areas are currently posted with signage and wording that are the result of CERCLA Records of Decision or interim corrective measures, but these postings are not consistent and, in some cases, do not indicate presence of a radiological hazard. These areas are not posted or controlled in accordance with 10 CFR 835, Occupational Radiation Protection.

3. Not all groundwater contamination has been fully and adequately characterized. While DOE has made extensive efforts to characterize the major sources and the extent of groundwater contamination and has established a water policy to ensure that public receptors are adequately protected, some areas have not been fully characterized. For example, sufficient data are lacking on the leading edges of both the Northeast and the Northwest Plumes. The density and positioning of monitoring wells are not adequate to assess the furthest movement or the discharge locations, such as streams, of the two northern plumes. The most recent plume map shows that movement has occurred under a portion of the Tennessee Valley Authority property, which borders the Ohio River.

4. Unclear assignment of responsibilities and lack of expertise have adversely impacted the understanding of environmental conditions. Neither DOE nor Bechtel Jacobs staff at the site have the requisite comprehensive knowledge of the nature of existing contamination in the various environmental media (surface water, sediment, soils, groundwater, and air). Sufficient technical personnel are not available to interpret the vast amounts of data associated with specific environmental disciplines.

5. Environmental information to the public has sometimes been delayed and is in forms not always clearly understood by the general public. Upon discovery of groundwater contamination in 1988, the site prepared a Community Relations Plan in response to CERCLA requirements. A review of current programs and activities to communicate information to the public identified a number of weaknesses, largely due to the lack of clearly defined roles and responsibilities for public communication. Annual environmental reports do not contain a clear summary of site conditions or public health risks. As a result, members of the public—including the Site Specific Advisory Board—have a perception that DOE does not adequately disclose information about hazards and risks.

Environmental Sampling Results

Environmental samples were collected and analyzed by the investigation team in an effort to confirm that the current analytical results being reported by the site are accurate and representative of environmental conditions. Site subcontractor personnel collected all the samples in accordance with approved procedures that follow EPA-established guidelines. The investigators witnessed the collection of all samples, and chain-of-custody forms were completed.

Groundwater samples were generally taken at the extremities of the reported plumes to confirm the extent of contaminant migration. Surface water samples were taken at major site outfalls flowing during the sampling period, and at points associated with surface waterways in the vicinity of the Plant. Soil and sediment were primarily sampled at outfalls and ditches near source areas of contamination. Groundwater, surface water, soil, and stream sediment were sampled and analyzed for key radionuclides and volatile organic compounds, including technecium-99, plutonium-239/240, neptunium-237, uranium-238, thorium-230, americium-241, and cesium-137, volatile organic compounds including trichlorethelene (TCE) and polychlorinated biphenyls (PCB).

Radiological and chemical contamination in groundwater, surface water, and soils/sediment were detected in some of the samples. With a few exceptions, the types and levels of contamination detected were consistent with the levels identified by past environmental monitoring conducted by the site, and do not pose a current public health or safety risk. The detailed results are provided and discussed in the investigation report.

Groundwater.—The oversight investigation team's groundwater sampling strategy involved sampling ahead of the plume in the direction of the plume movement in order to confirm the advance of the contamination. In a one-to-one comparison using previous data from the same wells, analytical results agreed with those in the site database and the chemical analyses of contaminants being reported by the site. Results indicate that the Northwest Plume is migrating northward through the TVA property.

Surface water.—Surface water samples were collected from nine selected locations along the Little and Big Bayou Creeks as well as at several Plant Outfalls where surface water was present. Radioactivity analyses for surface waters showed relatively low concentrations for all isotopes, with the North-South Diversion Ditch sample showing the highest levels of uranium and technetium-99. Transuranic and thorium isotopes were either not detected or were present in very low concentrations, consistent with prior sampling results conducted by the site. The surface water results are all well below the levels required in the DOE Order 5400.5.

Soil.—A total of eight soil/sediment locations were sampled for radionuclide and PCB contaminants adjacent to the site, and one was collected inside the site security fence near the Drum Mountain area. The magnitude of the radionuclide results was generally in keeping with historical data reported by the site.

Recommendations:

Radiological and chemical contamination from PGDP industrial activities have been released into the ground, soil, and air around the plant. These conditions have prompted DOE and regulatory organizations to take a number of steps to protect public health. Because of the limited duration of exposures of the public to contamination and the mitigation measures taken, DOE operations do not present a significant public health risk at this time.

Nevertheless, significant improvements in protection of the public and the environment are needed to avoid the possibility of a future health risk. Adequate funding and management emphasis on actual remediation activities are needed to address the sources of continuing contamination, to limit the degradation of contaminated buildings, and to control the continued spread of contamination pending cleanup. Exposure pathways need to be better characterized to fully document the technical basis and the site's conclusion that no significant public exposures to radiation sources, such as fugitive air emissions, are occurring. Site management also needs to improve the characterization of groundwater in several areas, such as the extent of progression of the Northwest Plume toward the Ohio River. Improvements in waste management practices are needed to address storage of materials in DMSAs and the degrading containers of low level waste.

Radiation Protection and Worker Safety and Health

The Bechtel Jacobs radiation protection program exists to protect individuals from radiological exposures that may occur as a result of DOE activities at the PGDP. These activities have changed during the 1990s as a result of the transition of gaseous diffusion operations to USEC. Despite the mission change, the nature, extent, and magnitude of contaminated facilities at the site present unique challenges, and highlight the importance and need for a comprehensive and robust radiological protection program.

During the early 1990s, radiological assessments, including the 1990 Tiger Team, identified fundamental program weaknesses in the site's ability to control potential exposures to transuranics and to conduct an effective contamination survey program. In response, the site initiated a number of improvements. While the investigation team identified similar deficiencies to those raised by the 1990 Tiger Team report, the magnitude in areas such as postings, procedures, air monitoring, and contamination control is less. Records indicate that the external doses to employees from the types of radiation present at Paducah are very low, and there have been no recent significant intakes of radioactive material.

The identified radiological protection problems are typical of a site that has had to cope with the same legacy hazards for many years and which is no longer in operational mode. There has been increasing informal reliance on worker knowledge rather than a disciplined and rigorous application of controls such as detailed radiation work permits, procedures, postings, barriers, and air monitoring. These deficiencies, while not significant individually, are of concern in the aggregate because of the uncharacterized hazards remaining, the unique and challenging risks associated with future hazardous cleanup, and the reliance on subcontractors who do not possess the historical knowledge of site radiological and contractor hazards, including transuranics, and the applicable precautions and controls. The identified weaknesses in radiological controls are exacerbated by a lack of DOE or Bechtel Jacobs oversight of radiation work practices.

Findings:

1. Radiological characterization of the workplace is incomplete, weakening the ability of the radiological control organization to identify hazards and institute controls necessary to ensure consistent and appropriate radiological protection for workers. There is a lack of knowledge as to the isotopic mix of radionuclides present in various work areas. This information has never been obtained through comprehensive characterization nor is it available in technical basis documentation. Radiological Control Technicians need this information to establish proper radiological controls. Procedures in place for planning and conducting radiological controls in the workplace presume knowledge of radiological control personnel about the isotopic mix in work areas.

2. There is a lack of rigor, formality and discipline in the development, maintenance, and implementation of the Bechtel Jacobs radiation protection program.

—Air sampler placement is not always consistent or adequate to sample the air in the work area or representative of the air breathed by the worker, and analysis of air samples is not timely. In many cases, the monitored work activity was already completed at the time final air sample activity was determined. Procedures do not identify those conditions that must be present to require isotopic analysis of air samples.

—Radiological surveys taken by Bechtel Jacobs in April and June 1999 concluded there was no need for dosimetry and radiological worker training for construction personnel working at the UF6 cylinder yard project. Subsequent dose rate measurements of the work area by the Investigation Team indicated that, based on an anticipated six-month job duration, worker doses would likely exceed the threshold for such controls, and workers should have been monitored and provided Radiation Worker I training. The finding led to a shutdown of work, radiological training for two workers, and the implementation of monitoring through use of dosimeters.

—Bechtel Jacobs cannot adequately demonstrate that the unconditional release of equipment from the site, such as the release of fluorine cells, is consistent with DOE requirements. While Bechtel Jacobs does have a procedure for unrestricted release of equipment, this procedure was not applied during the process of releasing the fluorine cells.

—Outdoor contamination areas, particularly in the vicinity of Drum Mountain, were not adequately posted and barricaded for the levels of radiological contamination present. Other onsite areas, primarily drainage ditches, were posted as contamination areas without specific information on the radiological or chemical hazards being present. Since there is no contamination monitoring of individuals leaving the site, there is the potential for contamination to be taken offsite.

It is important that DOE and Bechtel Jacobs recognize that the cumulative deficiencies, in what has the potential to be a viable and effective radiological protection program, warrant management attention. The contractor needs to establish rigor, a higher level of discipline and formality to protect worker health and safety during hazardous characterization and cleanup activities on-site. DOE and Bechtel Jacobs also need to improve oversight of subcontractor radiological safety and performance including accountability for adherence to applicable DOE requirements.

Worker Safety

Bechtel Jacobs has developed procedures for identifying, evaluating and controlling occupational hazards at PGDP and most have been identified. Completion of the cleanup mission at PGDP, however, will require a significant increase in activities involving the potential for hazardous materials exposure including the removal of buried waste and the inspection of the contents of thousands of drums of radioactive waste. This work involves the handling of material containing radioactive and chemical carcinogens, much of which has not been fully characterized. There have already been several occurrences of workers being contaminated as a result of drum handling and waste characterization activities. Many precursor conditions are developing that, if not addressed, will lead to decreased safety performance and an increased risk to workers.

Findings:

1. Criticality safety deficiencies in DMSAs pose an unnecessary hazard to workers in surrounding areas. Large amounts of legacy materials for which DOE is responsible are currently stored in 148 DMSAs across the site, including DMSA 'islands' within USEC spaces.

—These materials are not yet characterized, and 11 contain potential fissile material deposits and are identified as high priority. As a result, the risk of an inadvertent criticality is not known. Funding has not yet been provided to correct the deficiencies in all the DMSAs and eliminate the potential criticality safety hazard.

2. Safety and health procedures are not consistently applied and followed, and in some cases, hazards are not adequately addressed by those procedures.

—Of the occurrence reports submitted to DOE by Bechtel Jacobs since April 1998, a number were attributed to either inadequate procedures or a failure to follow procedures. For example, on May 27, 1999, it was determined that laboratory personnel working in a mobile field extraction laboratory had been exposed to methylene chloride above the 15-minute Short-Term Exposure Limit as defined by Occupational Safety and Health Administration regulation.

—The investigation team also observed that some safety and health procedures are not consistently followed. Sections of the site-wide procedure and the subcontractor's health and safety plan for confined space entry were not being followed at the L Cylinder Yard. Confined spaces were not evaluated, permitted, or posted in accordance with procedures. Sections of Bechtel Jacobs procedures on biological monitoring for industrial chemicals, and workplace air sampling were not being followed.

3. Bechtel Jacobs training programs do not ensure that all workers are knowledgeable of hazards and protection requirements, including those associated with transuranic contamination.

—The Bechtel Jacobs radiation safety training program does not include a process to assure that individuals have received the required training before working in controlled or radiological areas. Although required by procedure, mandatory training is not included in Radiological Work Permits.

—None of the current Bechtel Jacobs radiation safety training modules adequately addresses the presence of transuranic contaminants at the site. Transuranic training was provided once in 1992, and although DOE and Bechtel Jacobs' personnel believed that transuranic training was being conducted, in fact, the 1992 transuranic-based training was not incorporated into the ongoing radiation worker training program. Bechtel Jacobs Radiological Control Technician training does not include monitoring for transuranics, the release criteria to be used, or the use of isotopic analysis information to determine the need for controls.

DOE Line Oversight

DOE established a Paducah site office in 1989 to provide program direction and day-to-day oversight of contractor activities. DOE strengthened this oversight office in the early 1990s, in light of emerging environmental and worker safety issues such as the discovery of Technetium-99 in offsite wells and numerous sources of contamination contributing to a plume of contaminated groundwater.

With the final transition to NRC regulation of the enrichment facilities in 1997, the scope of DOE activities at PGDP decreased significantly. In April 1998, DOE transitioned from a management and operating contract with Lockheed Martin Energy Services to a management and integration contract with Bechtel Jacobs. The current level and effectiveness of line management oversight of environment, safety, and health and assurance of compliance with DOE requirements is a matter of concern. Programmatic deficiencies identified through the 1990s either continue or have recurred. Direction provided by DOE, primarily the Oak Ridge Operations Office in writing or verbally, regarding implementation of the management and integration contract has significantly reduced the level of oversight conducted by both the Paducah Site Office and Bechtel Jacobs. Consequently, programmatic problems have not been identified and corrected by line management.

Findings:

1. DOE has not conducted effective oversight of ES&H to ensure that Bechtel Jacobs and its subcontractors effectively implement all DOE and regulatory requirements.

—Oak Ridge has provided little written direction to the Paducah Site Office for oversight of the management and integrating contractor (Bechtel Jacobs). Written guidance stated that "the DOE role will center on establishing policies, standards, baselines, and objectives and measuring performance rather than focusing on day-to-day oversight and control." Consequently "day-to-day oversight" has received little attention.

—Neither Oak Ridge nor the Paducah Site Office has provided sufficient direction to Bechtel Jacobs to assure adequate oversight of subcontractors, although subcontractors are accomplishing an increasing amount of work.

2. Bechtel Jacobs has not conducted effective oversight of ES&H performance of its subcontractors to assure that subcontractors effectively implement DOE and regulatory requirements and are held accountable.

—Bechtel Jacobs' subcontractors do not consistently follow safety and health procedures. Numerous weaknesses were identified in the areas of procedure adherence, safe work practices, occupational medicine, and worker training. Some recent subcontractor work activities have resulted in unsafe work practices. Subcontractor prescreening by Bechtel Jacobs is not adequate to ensure the subcontractors have working programs in place that meet DOE requirements for Industrial Safety, Industrial Hygiene and Medical Surveillance.

—Although Bechtel Jacobs provides a measure of oversight of subcontractor training programs through quality assurance audits, surveillances, and readiness reviews, the oversight is not consistently applied and is at the discretion of the Bechtel Jacobs Project Manager.

—Planned reductions in staff within Bechtel Jacobs will further reduce Bechtel Jacobs' technical capability to conduct oversight and surveillance of subcontractor activities. Planned staffing changes include a reduction in Safety Advocates from four to one, and elimination of the training coordinator position. In addition, there are significant shortages in key safety disciplines, such as industrial hygienists.

Investigation Conclusions

There have been significant environment, safety and health improvements made at the Paducah site over the past ten years. Current operations do not present immediate risks to workers or the general public. At the same time, serious weaknesses remain in all major areas—environmental and public protection, worker safety and health, and DOE oversight that, in combination, undermine the confidence of workers and stakeholders and perpetuate the risks and hazards of legacy operations.

A key to regaining worker and public confidence, reducing hazards and risks to as low as reasonably achievable, and ensuring the continuing operation of the Paducah Plant, is to begin to accelerate progress in the cleanup effort, including compliance with impending initial major cleanup milestones including Drum Mountain and the waste buried beneath it. Systematic progress needs to be demonstrated in key cleanup and hazard reduction areas such as the elimination of the many sources of contamination, characterization and disposition of the DMSAs, the proper storage and shipment off-site of low level waste, and the removal of hazards and proper upkeep or demolition of shutdown hazardous facilities. Other areas where timely improvement is needed include:

—Establishing a high level of discipline and rigor in the radiological protection program and other programs affecting worker safety, such as criticality safety. Programs should include verbatim compliance with posting and barrier requirements, improved radiation work permits, comprehensive radiological training, strict procedure use and compliance, characterization of materials to improve effective hazards analysis, and the use of engineered hazard controls whenever possible.

—Strengthening communications and outreach to workers, the public, and the stakeholders to ensure understanding, confidence in site operations, and empowerment in contributing to cleanup strategies, priorities, and decisions. This is particularly important for the Site Specific Advisory Board whose charter is to contribute to site cleanup through involvement in establishing priorities and milestones and achieving public support.

—Improving DOE and contractor oversight of ES&H performance to ensure adequate subcontractor safety performance, accountability for compliance with DOE requirements, and continuous improvement.

Continued improvements in safety management will be particularly important as the Paducah Site initiates additional site cleanup and remediation activities. Such work presents unique hazards (e.g., handling material containing radioactive and chemical carcinogens that has not been fully characterized) and has already resulted in several occurrences of workers being contaminated in the limited remediation efforts to date. The need for effective safety management is further highlighted by the fact that, under the managing and integrating contractor concept, a large fraction of the potentially hazardous work will be performed by subcontractor employees, some of whom do not have a historical knowledge of site hazards or controls. As subcontractor cleanup and waste management activities increase, increased surveillance and oversight will be needed by Bechtel Jacobs and DOE personnel who are knowledgeable of DOE requirements.

OTHER PADUCAH-RELATED ACTIVITIES

Determine Flow of Recycled Materials through the DOE Complex.—DOE and its predecessor agencies produced more than 100,000 metric tons of recycled feed or waste streams containing trace quantities of fission products and plutonium. This material was sent not only to Paducah, but also to other sites in the DOE complex. Today, our understanding of where that material went is limited. Secretary Richardson requested a study that would provide a clear understanding of the flow and characteristics of this recycled material. DOE is concerned not only with the flow of this material, but also its characteristics such as the level of residual plutonium and fission products. The mass flow project will address the flow and characteristics of recycled uranium over the last fifty years. We expect this work to be complete by June 2000. The specific goals are to:

—Identify the mass flow of recycled uranium throughout the DOE complex from early production to mid-1999 and create a publicly available unclassified inter-site flowsheet.

—Identify the characteristics of, and contaminants in, the major uranium streams, including the technetium, neptunium, plutonium or other radioactive content of concern to worker or public health and safety.

—Conduct site mass balance activities to identify any significant concern for potential personnel exposure or environmental contamination.

Worker Exposure Assessment Project.—Secretary Richardson has committed to fully address health concerns of current and former Paducah workers, especially where records are less than complete, or where worker exposure to plutonium and other materials has not been well characterized. To address this gap, an aggressive and exhaustive search of records is being conducted at Paducah for the time period ranging from the early 1950s to the present. Assessments will then be performed by analyzing the exposure records of current and former workers to determine the extent and nature of exposures, focusing on exposure to transuranics. The activity will include identifying, retrieving and reviewing exposure records. Should records prove to be poor or non-existent, DOE would perform detailed reviews of relevant plant process and monitoring data as well as extrapolations based on available exposure information.

The goal of this effort is to establish the potential ranges of worker radiation exposures and identify, document and communicate the radiological issues that may have affected worker health at the Paducah site since its opening. This work will help inform Paducah workers of their potential radiation exposure and will help determine whether there may be any potential for adverse worker health impacts from radiation exposure while working at the Paducah plant. We are currently investigating the conduct and cost implications of similar exposure assessments at Portsmouth and Oak Ridge.

CONCLUSION

Finally, Mr. Chairman, I want to emphasize that Secretary Richardson, on behalf of the entire Administration, takes the concerns that have been raised seriously and is committed to investigate and resolve them. The investigation is both independent and comprehensive. As you have seen, it has already begun to serve to get out the facts and correct any current safety shortcomings. The existing environmental compliance agreement that guides remedial actions and schedules at the site has been agreed to by DOE, the State of Kentucky and the Environmental Protection Agency. Where the investigation team's initial observations suggest that modifications to this agreement, including adjustments in priorities, may be warranted to protect the public and worker health and safety, we will pursue them.

We need to determine how well the workers knew of and understood the hazards they were working with, and how well they were protected from these hazards—even in very small amounts. We will learn much more as our investigation moves ahead and seeks to confirm—in today's regulatory environment—whether the presence of these materials represented a potential health risk at Paducah or any other DOE plant.

We will continue our efforts in a manner that is forthright and responsive to the public's need for timely information, while at the same time being careful that our answers are correct. We will also continue to work in a cooperative and transparent way with the workers, their representatives, the public, and the Congress. Secretary Richardson has made it clear that the days of secrecy and hiding information are over. We are committed to getting accurate information and doing so in a responsible manner. We are also committed to treat our workers dignity and with fairness.

Thank you for the opportunity to testify. I would be happy to answer questions from any of the Subcommittee members.

WORKERS' COMPENSATION PROGRAM

Senator MCCONNELL. Now Dr. Michaels, you heard David Fuller's testimony that DOE and its predecessor the Atomic Energy Commission aggressively fought all workers' claims of occupational illness and deliberately withheld information for fear that it might result in higher compensation to workers. Add to that the horrific treatment by the Department of Energy of Joe Harding and his family. What is your office doing to rectify this situation?

Dr. MICHAELS. Yes, sir. The day that I was sworn in, Secretary Richardson asked me to go to Oak Ridge and speak with the workers who Dr. Bird talked to you about. Secretary Richardson instructed me to listen to them and find a way to help. I have since been across the complex at the request of Secretary Richardson and found similar situations to what you have described.

What we are now doing is working on a workers' compensation program that will do exactly what the president of the local union, David Fuller, described. We are trying to come up with a program, and the administration has announced its support for a program around beryllium disease, to do exactly this, to provide an alternative to State workers' compensation benefits to workers in the

DOE complex who put their lives on the line making materials for nuclear weapons.

In July President Clinton signed a memo calling for interagency review of occupational illness across the DOE complex. That should be done by March and we expect to have proposed legislation to address these issues some time after that.

INDEPENDENT REGULATORY OVERSIGHT

Senator MCCONNELL. Your phase one assessment found that DOE continues to make the same regulatory errors that were identified in the 1990 DOE tiger team report. Considering the Department's proven inability to serve as both regulator and site cleanup manager, is it not time that we move the oversight responsibilities from DOE to an independent regulatory body?

Dr. MICHAELS. Sir, I do not think that is necessarily the answer. I think we saw the same problems that the 1990 tiger team saw, but not in the same dimension or magnitude. We found great progress had been made. Certainly there has been some major backsliding.

The problems facing the DOE complex are the most technically complex problems facing any work place in America today, the legacy of the nuclear weapons production. While there are some advantages to bringing in outside regulation, it is not clear that that alone will make the difference. I think we have to look at good ways to investigate and do regular oversight. I think our office and the independent regulators at DOE have a major role to play in that.

PLUTONIUM AND URANIUM CONTAMINATION

Senator MCCONNELL. The claims of plutonium contamination of the creek beds is really troubling. As the regulator at the Paducah plant, how do you explain the significant radiological contamination outside the fence, how did it get there, and how long has it been there?

Dr. MICHAELS. Sir, the source of the plutonium we believe is the contaminated radioactive uranium tailings, essentially, that were brought in, uranium feed that was used in the Paducah system. There was plutonium, neptunium, and other materials that then were released and continue to be released in the outflows.

They are a significant problem. We believe that the levels are quite low, but we believe that no exposure is a good exposure or a safe exposure and we should be doing everything we can to reduce and eliminate that exposure.

Senator MCCONNELL. You heard David Fuller testify that workers were taking the contaminated uranium dust home with them and reported that many workers acknowledged that they would frequently find green uranium dust in their clothes and even in their bed sheets. How do you think that something like this could occur and does it pose any risk to the workers' families?

Dr. MICHAELS. This is one of the questions we are going to be looking at in the second phase of the study, where we are looking at exposures before 1990. Certainly I am concerned about it. It sounds as if the radiation control in the plant during that period was lacking a lot of the fundamental things we would expect to see.

I hope to be able to give you an answer to that some time in January.

Senator MCCONNELL. According to experts, it is absolutely critical that the Department reconstruct the radiation exposures that workers might have been exposed to in an effort to determine future health risks. What is DOE doing to provide this information and are you confident that DOE will be able to develop an accurate picture of what the work force was exposed to between 1950 and 1970, for example?

Dr. MICHAELS. What DOE is doing at Paducah, as well as Portsmouth and Oak Ridge, is we are attempting to bring in a reliable, independent outside group, in this case the University of Utah, to recreate the doses that people got across the three gaseous diffusion plants. This is being done under our joint aegis with the PACE union, and PACE has a health physicist who is overseeing it at the same time that we are, so that everybody is confident when we get the results back that this was done as well as it possibly could be done.

NEW AREAS OF CONTAMINATION

Senator MCCONNELL. How many new areas in and around the plant have your teams roped off or identified as being inadequately marked since the investigation began?

Dr. MICHAELS. I am going to ask Bill Eckroade from our staff to help address this. Bill was the director of the environmental component of phase one.

Mr. ECKROADE. There were three areas of particular concern to the investigation team in the vicinity of the plant security fence, those being Outfall 11, Outfall 15, and the North-South Diversion Ditch. While it was known that contamination was present from historical sampling at all those locations, sampling from the team's efforts found elevated levels of a variety of isotopes that had not been previously detected.

Senator MCCONNELL. Are there any areas that now require a higher level of radiation training to assess than was required prior to your investigation?

Mr. ECKROADE. In one outfall, Outfall 11, when we went to do the sampling, that area has historically been accessed by sampling crews to take measurements. Upon sampling by the team, the site had sent a certified technician to take readings, identify the elevated readings through a scan, and then changed the entry requirements to require additional protective equipment, and has subsequently posted that area requiring additional protection for entry.

Senator MCCONNELL. Dr. Michaels, will all your studies of the DOE facility be properly reviewed by an independent entity to ensure objectivity and accuracy?

Dr. MICHAELS. That is an interesting question, Senator. Our investigation——

Senator MCCONNELL. I am waiting for an interesting answer.

Dr. MICHAELS. I had not considered that. These investigation reports or oversight reports we do not send out for independent peer review because they are investigative reports rather than scientific conclusions. On the other hand, if we do epidemiology, for example, we will definitely have that peer reviewed. But some of our investigations really I do not think warrant peer review.

Senator MCCONNELL. Why do you suppose the DOE test did not pick up on the exposure levels to which Joe Harding had obviously been subjected? Is it possible the Department was testing for the wrong thing or that somehow Mr. Harding's results were falsified?

Dr. MICHAELS. That is a very good question that we will attempt to answer in phase two. I do not know what the Department was examining before 1990 because we have not yet collected that data, but we will be looking exactly at that. We note, though, that before 1990 the health physics program at Paducah was lacking, and we will look to see whether they were measuring anything or just the wrong thing.

Senator MCCONNELL. Could you explain to the committee the results of your recent soil sampling in the area of the Little Bayou Creek and the Big Bayou Creek? Was there any contamination? What about plutonium or tech-99, and if there was any of that how do you suppose the material got there?

Dr. MICHAELS. Let me bring up Bill Eckroade again.

Mr. ECKROADE. The sampling that the investigation team did with respect to sediments, we had taken nine samples, primarily in the vicinity of the plant location, several down the reaches of the Big and Little Bayou. Contamination was primarily identified in the locations in the near vicinity of the site, essentially the outfalls that I mentioned before, the K–15, the K–11, and the North-South Diversion Ditch.

Basically, the contamination is the result of historic operations from past enrichment operations discharged through various mechanisms into the environment, and it continues to spread in those receptors.

Senator MCCONNELL. What is it, plutonium, tech-99, what?

Mr. ECKROADE. In our report, we actually list a table of different isotopes that we detected. We detected plutonium 239, some levels of cesium, in particularly Outfall 15, and tech-99 at lower levels.

Senator MCCONNELL. How do you suppose it got there?

Mr. ECKROADE. Past operations. Certainly parts of the enrichment processes concentrated some of the impurities and they were subsequently released into the environment.

Senator MCCONNELL. Finally, Dr. Michaels, are you aware of the Senate Government Affairs Committee report produced in December 1989 that identified irresponsible behavior on the part of the Atomic Energy Commission that contributed to the radiation exposure of workers at Federal facilities? And if you are, what specific reports came out of this report and were any of those reforms implemented at Paducah?

Dr. MICHAELS. I am not familiar with that specific report, no, sir.

Senator MCCONNELL. Well, you might want to do that.

Dr. MICHAELS. I definitely will. In fact, there is a staffer who served on that committee in this room today and I will ask him for it.

Senator MCCONNELL. Apparently that report came out in December 1989, almost 10 years ago.

Well, Dr. Michaels, you have your work cut out for you and we are looking forward to hearing from you periodically. We would love to see some tangible progress made.

Dr. MICHAELS. Thank you, sir.

Senator MCCONNELL. Thank you.

STATEMENT OF CAROLYN L. HUNTOON, ASSISTANT SECRETARY FOR ENVIRONMENTAL MANAGEMENT, DEPARTMENT OF ENERGY

ACCOMPANIED BY BILL MAGWOOD, ASSISTANT SECRETARY FOR NUCLEAR ENERGY, DEPARTMENT OF ENERGY

Senator MCCONNELL. The next panel and final panel: Carolyn Huntoon, Assistant Secretary of DOE, Office of Environmental Management; Bill Magwood, Director of the DOE Office of Nuclear Energy, who will not have a statement, but will just be available for questions; Secretary Bickford of the Kentucky Department of Natural Resources; and Richard Greene, EPA Region IV. EPA works with the State and DOE on the cleanup issues.

Let me remind the witnesses again that 5 minutes means 5 minutes. Dr. Huntoon, why do you not start.

Dr. HUNTOON. Thank you, Senator McConnell. I am here today to tell you about the cleanup program at Paducah and what we intend to do to correct the program's shortcomings that have been identified.

ENVIRONMENTAL LEGACY OF PAST WEAPON PRODUCTION

In the 3 months I have been at DOE, I have visited 8 of our sites around the country, including Paducah. The enormity of the environmental legacy from building the nuclear weapons is evident at every one of these sites. Everywhere I have gone I have seen evidence of cleanup challenges we face, and at Paducah I saw the famous Drum Mountain. I saw scraps of metal. I saw buildings like C–400, where most of the TCE now contaminating the groundwater was used. I talked to workers on the site. I talked to people out doing the remediation.

After seeing the site and reading the report, I agree with you and the local residents that the cleanup should proceed as expeditiously and safely as possible. I recognize the magnitude of this challenge both at Paducah and across the entire DOE complex.

The reality is we have neither the money nor the technology to clean up as quickly as any of us would like to do. At Paducah we have spent about $388 million implementing a three-pronged cleanup strategy to address the site's environmental contamination. It is a strategy we have developed and periodically re-evaluate with the Commonwealth of Kentucky and the EPA regulators, with input from local citizens.

First, we are addressing the most urgent risks, particularly threats to the public from offsite contaminations. Since 1988 we have been ensuring that local residents have safe drinking water, first by providing bottled water, then by providing a permanent solution by paying for municipal water hookups for over 100 residents. We continue to pay the water bills for these homes.

Second, we are working to identify the nature and scope and location of the contamination, which involves characterizing and controlling the hot spots and other suspected sources of these offsite contaminations. For example, we have drilled 340 wells, of which we routinely monitor 165 to define the groundwater plume. We have constructed two groundwater treatment systems and have treated 600 million gallons of contaminated groundwater. We have constructed an onsite sanitary landfill and disposed of 14,000 tons of solid waste, and we have disposed of 5.8 million pounds of hazardous and radioactive waste.

LONG-TERM CLEANUP SOLUTIONS

Third, we are working on long-term cleanup solutions. I understand and share everyone's concern that we have not moved ahead fast enough. However, the work to determine and characterize the nature, extent, and the source of the contamination both on and off site was a critical precursor to being able to move ahead with the solutions that can now be implemented safely for the workers and the environment.

The Congressional cuts in 1998 and 1999 for the uranium D&D fund which pays for this work further slowed our cleanup efforts. I want to thank you personally and the work of the rest of the Kentucky delegation for their efforts this past year in securing more funds. I know that with Paducah's share of the $10 million additional money we are going to be able to initiate and complete the removal of Drum Mountain by the calendar year 2000.

Let me turn back to the concerns that were raised by last week's report. I want to assure you that Secretary Richardson and I take this very seriously. I have read it. I think it is fair. It is a fair assessment. We need to correct these conditions.

We have completed or initiated a number of actions, including: Posting new signs for radioactively contaminated areas on DOE property that previously had only warnings; we started a top to bottom review of the radiation control program at Paducah as well as the two gaseous diffusion plants. The review at Paducah, which started on October 18, will be completed by mid-November, with a report due to me by November 30.

We have increased contractor oversight by assigning more Federal employees to Paducah whose sole job will be to monitor the safety of our facilities and operations. We have sent two employees to the site on a temporary basis until we can fill these jobs permanently. The first permanent employee will be there on November 7.

Using the $6 million in additional money for fiscal year 2000 funding to accelerate the removal of Drum Mountain, we expect to complete the work by the end of the calendar year, pending approval of the Commonwealth of Kentucky and the EPA regulators. We are also evaluating other opportunities to accelerate work at Paducah should additional funds become available.

I have deployed a technology assessment team from our Idaho lab which specializes in subsurface water problems. They will give me their report by November 30 on things that they believe that can be done immediately to help us with the groundwater situation.

We are sampling the roofs of the shut-down buildings to see if they are emission sources. We will have that data, pending approval of the Commonwealth of Kentucky, this January.

Reviewing how we communicate with the public, we have set a plan for improved communications. That plan will be put in place by November 9.

Sitting down with the Commonwealth of Kentucky and EPA regulators on November 8 and 9, we will review our existing cleanup agreements and priorities fundings and modify those as necessary.

PREPARED STATEMENT

I have directed the site to develop a more complete plan that will address the remaining findings. I have also directed Oak Ridge operations office to ensure they apply the lessons learned to review the other gaseous diffusion sites at Portsmouth and Oak Ridge.

[The statement follows:]

PREPARED STATEMENT OF DR. CAROLYN L. HUNTOON

Thank you, Mr. Chairman. I appreciate the opportunity to bring the Subcommittee up to date on the Department of Energy's environmental cleanup program at the Paducah Gaseous Diffusion Plant in Kentucky.

In the nearly four months I have been at DOE, I have visited eight sites around the country, including Paducah. I have seen for myself the contamination problems at Paducah—the infamous Drum Mountain with its thousands of crushed drums, the scrap metal piles, the buildings and areas that are the source of much of the contamination in groundwater. What has impressed me the most is the enormity and the complexity of the legacy of environmental problems left behind from our nuclear weapons work.

My goal at the Paducah Gaseous Diffusion Plant site is to complete cleanup of the site as expeditiously and cost-effectively as possible. I want to accelerate cleanup to reduce risks and costs in a manner consistent with my strong commitment to the safety of workers, the general public, and protection of the environment. I want to be sure that we are addressing site contamination problems in the right priority. We will continue to work in close partnership with the Commonwealth of Kentucky and the U.S. Environmental Protection Agency (EPA), workers, and local citizens at the site on all aspects of the cleanup, including setting cleanup priorities. We will need to deploy innovative technologies and streamline the regulatory review process to maximize dollars we spend on actual cleanup. I will work with the Congress to seek the necessary funding to complete cleanup by fiscal year 2012.

In my statement to you today, I will provide you with an overview of the Environmental Management program and the cleanup challenges at the Paducah site, describe the strategy which has resulted in the most immediate risks at the site being addressed, explain our progress and plans to address longer-term contamination problems, and finally discuss the funding profile and issues at Paducah. Before I move to the specifics of our cleanup work, however, I would first like to talk about my commitment to safety and our efforts to ensure that the health and safety of the workers are protected during the cleanup work process.

ENSURING HEALTH AND SAFETY

Recognizing this Committee's interest in working conditions at the plant, I wanted to assure you that my first priority as Assistant Secretary for the Environmental Management (EM) program is safety—safety of the contractor and Federal workers that run our facilities and of the public in the communities around our sites. Since being confirmed as Assistant Secretary last July, I have established principles that will govern implementation of the program. Safety of the workers and the public is paramount, and I will hold my managers accountable for ensuring the workers and the public are protected.

We are working to ensure that cleanup activities at Paducah are conducted in a manner that protects the health and safety of the workers and the public. In response to the review by the Office of Environment, Safety and Health, the Management and Integrating contractor at Oak Ridge, Bechtel Jacobs Company, which manages cleanup at the three gaseous diffusion plants in Portsmouth, Ohio, Oak Ridge, Tennessee, and Paducah, is undertaking an independent, top-to-bottom review of its radiation control programs at the sites to ensure the controls and procedures in place are in compliance with DOE requirements and are being fully implemented. Their review at Paducah began earlier this month, and the assessment of the other sites will be completed by mid-November. The results of the reviews will be available by the end of November. I assure you that, should the review identify any gaps or areas that need improvement, we will work with the contractor to see that the necessary changes are made to ensure we are protecting the workers who are carrying out the cleanup work, while also protecting the public and environment.

THE ENVIRONMENTAL MANAGEMENT PROGRAM AT PADUCAH

The 3,500 acre site in Paducah—including 750 acres within the fenced security area and 2,000 acres leased to the Kentucky Department of Fish and Wildlife—is among the Department's smaller sites. The site is still producing enriched uranium for commercial nuclear reactors. The enrichment operations were privatized in 1993 under the auspices of the U.S. Enrichment Corporation (USEC). USEC is responsible for all primary process facilities and auxiliary facilities associated with the enrichment services and for waste generated by current operations. The Department has responsibility for facilities, materials, and equipment not needed by USEC for their operations. The cleanup of environmental contamination at the site and management of legacy waste is DOE's responsibility. The Department will ultimately have primary responsibility for deactivation and decommissioning of the plant when operations cease, just as it now does for the former gaseous diffusion plant at Oak Ridge.

Within the Department, the Office of Environmental Management and the Office of Nuclear Energy, Science and Technology share responsibility for different aspects of the management and cleanup of the site. Nuclear Energy is the site "landlord." It has responsibility for administering the lease of facilities to USEC, storage and maintenance of the cylinders containing depleted uranium hexafluoride at the site, and other landlord functions such as maintenance of roads and fences outside the security area. Nuclear Energy is responsible for surveillance and maintenance of surplus facilities not leased to USEC.

The Office of Environmental Management (EM) bears primary responsibility for cleanup. This includes remediation of environmental contamination caused by releases of hazardous and radioactive materials into the environment from previous operations and disposal practices. We also are responsible for management and disposition of "legacy" waste generated by operations before USEC assumed ownership, as well as scrap metals stored on-site. EM also conducts surveillance and maintenance for two site plants, including ancillary buildings associated with the plants, that have been shut down—the C–410 Feed Materials Plant and the C–340 Metal Reduction Plant—to control releases from the buildings.

The cleanup problems and contaminants we face at Paducah are diverse, and include both on-site and off-site contamination from radioactive and hazardous materials. The environmental problem receiving our earliest and most focused attention has been groundwater contamination, which has contaminated private residential wells. The contaminants are traveling in two plumes in a northeasterly and northwesterly direction, extending off-site approximately one and a half miles toward the Ohio River. We have also recently discovered a smaller plume moving to the southwest that appears not to extent beyond the site boundaries. The primary contaminants in the three groundwater plumes are trichloroethylene (TCE) and technetium-99. TCE is an industrial degreasing solvent which was used in large quantities from the early 1960's until 1993 to decontaminate equipment and waste material before disposal. Because of widespread industrial use, TCE is a very common contaminant in groundwater at DOE sites and at private sector and Federal facility sites across the country. Technetium-99 is a beta-emitting radionuclide and a fission by-product, introduced into the plant as part of the Reactor Tails Enrichment Program that ran from 1953 to 1975.

There are also numerous contaminated areas around the site where chemical wastes, such as polychlorinated biphenyls (PCBs) used in electrical transformers; radioactive wastes; or trace amounts of plutonium and other transuranics (elements with atomic numbers greater than uranium), were disposed or inadvertently spilled or otherwise released to the environment. For example, significant quantities of TCE got into the environment from leaky sewers and from the ventilation system. Contamination has migrated to or threatens surrounding soils, groundwater, creeks and ditches. We also must safely manage and disposition about 60,000 tons of scrap metal, and 6,000 cubic meters of low level waste in drums, much of which is currently stored outdoors and exposed to the elements.

Cleanup of the Paducah site continues to be carried out under the direction of Federal and State regulatory agencies. The first regulatory vehicle was a consent order with EPA issued in 1988 to cover initial groundwater measures to address drinking well contamination and characterization of the plumes. The Paducah site was listed on Superfund's National Priorities List in 1994 and, in 1998, DOE, the Commonwealth of Kentucky and EPA signed a Federal Facilities Agreement that provides the framework for cleanup, establishes priorities and enforceable milestones, and integrates cleanup requirements. We carry out our work in accordance with this agreement and other environmental laws, as well as the Atomic Energy Act (AEA), and DOE rules and orders that implement the Department's AEA responsibilities for managing radioactive materials.

Beginning with door-to-door outreach to local residents when contamination was first discovered in residential wells in 1988, the Department continues to work with the local community to provide information and hear their concerns on contamination problems and the site cleanup actions and priorities. DOE has held periodic public meetings since 1989 to keep residents informed of contamination problems and cleanup progress. It has also supported several advisory groups, including a Neighborhood Council of plant neighbors that provided input to DOE, and later to USEC, in the early 1990s. The Site Specific Advisory Board, formed in 1996, now serves as a primary vehicle for two-way communication on the cleanup with the local community.

CLEANUP ACTIONS TO DATE: THE MOST IMMEDIATE OFF-SITE THREATS HAVE BEEN ADDRESSED

Our strategy is risk-driven. Our highest priority has been to address the most immediate threats to the public from off-site contamination. We have also focused on identifying and eliminating the "hot spots" and other suspected sources of off-site contamination. And we have worked to characterize the site and analyze solutions to develop a sound technical basis for long-term action and ensure our workers doing the cleanup are safe. This strategy and our priorities for action have been developed in conjunction with our State and EPA regulators and others with concerns at the site, and are incorporated into our cleanup agreements. With the State and EPA, we have worked to set priorities for the available funding each year to ensure it is used to address the highest risks and to support long-term cleanup.

We have successfully completed actions to address the most immediate off-site risk, specifically the threat posed by the contamination of off-site residential wells from contaminated groundwater. Upon discovery of contaminated wells near the Paducah plant in 1988, the Department immediately provided bottled water to the residents whose wells were contaminated and began sampling nearby residential wells and monitoring wells to determine the extent of contamination, ultimately sampling about 400 off-site wells. The sampling results indicated TCE concentrations in six residential wells were greater than the EPA drinking water standards of five parts per billion. The Department put in place a residential well sampling program, and entered into an Administrative Consent Order with EPA to thoroughly investigate the source of contamination and take appropriate actions.

After completing the groundwater investigations, the Department, working with the municipal authorities, funded the extension of 12 miles of municipal water supply line to over 100 residences and businesses whose wells were contaminated. We are also paying their water bills. Through our characterization efforts, the Department has also identified the areas of the plumes with the highest concentrations of contaminants and has installed groundwater pump and treat systems in each plume to contain the spread and treat these higher contaminant concentrations. These treatment systems, installed in the Northwest plume in 1995 and in the Northeast plume in 1997, have already treated about 600 million gallons of contaminated groundwater. Monitoring data show that these systems are successfully containing the spread of these high concentration areas.

While we have addressed the most urgent risk to the public from the groundwater plumes, we continue to sample groundwater on a routine basis using a monitoring network of some 165 residential and other wells installed to track contaminant migration.

WE HAVE TAKEN INTERIM ACTIONS TO MITIGATE OFF-SITE CONTAMINATION SOURCES

The second prong of our cleanup strategy has been to characterize the contamination and control "hot-spots" and other suspected sources of off-site contamination. We have made progress with these efforts. We have:

—removed 162 cubic yards of contaminated soil from areas that have high concentrations of contaminants;

—taken several steps to reduce potential contamination associated with the North-South Diversion Ditch, where the highest levels of plutonium and uranium were found. We have installed a treatment system for effluents from the

C–400 Cleaning Building to reduce concentrations before discharge, and have installed an approximately 1300-foot piping system that bypasses about half the length of the ditch to reduce the potential for sediment contamination;
—closed 9 leaking underground storage tanks that stored petroleum products or cleaning solvents which were found to be contaminating soils and potentially groundwater;
—excavated about 60 cubic yards of contaminated soils from a concrete rubble pile located in the Ballard County Wildlife Area;
—installed an impermeable cap over the uranium burial ground and enhanced the existing cap on a sanitary landfill to reduce leachate migration from rainfall infiltration;
—closed on-site low-level waste burial grounds and waste storage areas; installed sediment controls at the scrap yards and drainage ditches to mitigate surface water and sediment runoff; and
—installed institutional controls for off-site contamination in surface water, outfalls, and lagoons.

While we have not addressed all potential sources of groundwater and soil contamination, we have eliminated the contamination "hot spots" that have been identified to date through our characterization efforts, and have mitigated other key potential sources of off-site contamination.

PROGRESS AND PLANS TO ADDRESS LONGER-TERM THREATS

Most of our "on-the-ground" cleanup actions to date have been directed toward eliminating immediate risks and contamination hot spots, particularly those contributing to off-site contamination. We have, for the most part, accomplished that objective, and site priorities are now shifting to cleanup of on-site sources contributing to groundwater and surface water contamination, and to long-term cleanup remedies.

In addition, like any other complex cleanup project, much of our work to date has been directed toward the characterization and assessment of the contamination at the site, providing the information necessary to identify and prioritize cleanup activities and to devise sound technical solutions. While less dramatic than actual cleanup, this work is a critical step in cleanup. Because of the hazardous nature of the contaminants and the processes involved in cleanup, characterization is also a critical step in protecting the workers who are doing the cleanup. Although there is more characterization and analysis to be done, our efforts will increasingly shift to actual cleanup, making use of the data and information that has been developed.

In the fall of 1998, after receiving the reduced fiscal year 1999 appropriation, the Department and the State and EPA regulators reviewed the cleanup strategy to ensure we were making the best and most efficient use of our resources. The result was a revised approach incorporated into the agreement. Rather than requiring separate evaluations and decisions for some 30 individual sources, the cleanup work now is organized around four contaminant pathways, specifically the groundwater plumes, creeks, burial grounds, and surface soils. This new approach enables us to better integrate our efforts and to streamline the administrative process by significantly reducing the number of individual remedy selection decisions needed, a lengthy process that can take as much as two years; this will allow us to focus more resources on cleanup.

The activities planned for fiscal year 2000 illustrate the shift from the focus on immediate risks and interim actions to the next phase of cleanup. Our groundwater cleanup activities this fiscal year include the start of operation of an innovative treatment technology, referred to as the "Lasagna" technology, to treat TCE-contaminated soil. Named for the layers of sand, silt and clay beneath ground level, the Lasagna process generates an electric field and uses chemical means to destroy the TCE. We will use this technology to remediate shallow soils in the former Cylinder Drop Test Area, a major source of TCE contamination in groundwater; we expect to complete TCE removal in fiscal year 2001. We will also conduct a treatability study for the Southwest plume for evaluate an innovative in-situ groundwater technology, and will continue to make progress on evaluation and selection of a final remedy for the groundwater plumes.

One of my priorities is to bring the best science and technology to bear on solving the cleanup challenges facing the Department. I am forming a Technology Deployment Assistance Team at Headquarters to help the sites identify innovative technologies that can solve cleanup problems in a more efficient and less costly manner. I plan to couple this effort with ongoing efforts to accelerate technology deployments across DOE sites. I know the groundwater issues at Paducah can benefit from an additional science focus and have directed a Technology Deployment Assistance

Team that will include scientists from the Idaho National Engineering and Environmental Laboratory to work with the Paducah site to identify innovative technologies for characterizing, monitoring, and remediating groundwater plumes. They are to report their findings to me by November 30.

Our planned activities in fiscal year 2000 to address surface water contamination include accelerating the removal and disposal of "Drum Mountain," a large scrap pile containing thousands of drums, which is a suspected source of contamination of the Big and Little Bayou Creeks from surface run-off. The additional funds provided by Congress in the fiscal year 2000 appropriation will enable us to complete the removal of the drums by the end of next year, a year earlier than previously planned. This project will allow us to remove a major impediment to cleaning up the burial grounds as well as eliminate a potential contamination source. We will also continue characterization of other source areas draining into these creeks this year.

Our current schedule anticipates completion of cleanup at Paducah in fiscal year 2012, with long-term stewardship to monitor the site and ensure the remedies remain protective required after that. Based on the current schedules in the agreement with the State and EPA, we plan to issue a Record of Decision selecting the remedy for TCE and technicium-99 contamination in groundwater in fiscal year 2001 based on the evaluation of a number of innovative technologies, and begin implementing the remedy the following year. We also plan to complete work on surface water, surface soils and burial areas by 2012, including removal by 2003 of 60,000 tons of scrap metal stored in piles. Finally, we will complete removal of all mixed and low-level legacy waste by 2012 by shipping the waste offsite for treatment and disposal.

The cost of active cleanup at Paducah through 2012 is estimated to be approximately $700 million. There will be additional costs associated both with long-term monitoring and maintenance of the cleanup and decontamination and decommissioning of the uranium enrichment process buildings and other buildings at Paducah. .

FUNDING THE CLEANUP OF THE PADUCAH SITE

Cleanup activities for Paducah are funded through a separate account, the Uranium Enrichment Decontamination and Decommissioning Fund, which also funds cleanup at the Portsmouth Gaseous Diffusion Plant in Ohio, and at the gaseous diffusion plant in Oak Ridge, Tennessee, now called the East Tennessee Technology Park (ETTP), which ceased uranium enrichment operations in 1985. The fiscal year 2000 appropriation for the Uranium Enrichment Decontamination and Decommissioning Fund is $250 million, of which $220 million supports cleanup of the three gaseous diffusion plants. Cleanup of the Paducah site received about $36 million in fiscal year 1999 and $43.5 million in fiscal year 2000. This is comparable to the level of funding provided for cleanup at Portsmouth. The funding in fiscal year 2000 includes $6 million from the additional funding appropriated for cleanup activities at the gaseous diffusion plants in response to the budget amendment.

Funding for EM activities at the Tennessee facility is significantly higher, reflecting the deactivation and decontamination of the process buildings at the site and the excess materials and equipment in the buildings—facilities which are still in operation at Paducah and Portsmouth. EM also funds landlord operations at ETTP, costs that are currently covered by USEC and the Office of Nuclear Energy at Portsmouth and Paducah. Landlord responsibilities at ETTP accounted for about 25 percent, or $41 million, of the budget for ETTP in fiscal year 2000. There are also additional waste management facilities funded at ETTP, including the TSCA incinerator, the only low-level waste treatment facility in the DOE complex with permits to treat radioactive waste that also contains hazardous or PCB-contaminated waste.

The EM program has invested approximately $400 million in the cleanup of Paducah from fiscal year 1988 through fiscal year 1999. It is important to understand, however, that not all of those funds support visible cleanup. Like other sites in the complex, a significant portion—more than a third at Paducah—goes simply to "keeping the doors open" and maintaining minimum safety conditions at the site. It includes activities such as maintaining safe storage of about 50,000 drums of legacy waste, surveillance and maintenance of the shut down facilities, operation of a solid waste landfill, routine monitoring of groundwater wells, and program management support. Another significant portion of these funds, again about a third, have been used for characterization and assessment of the site, a critical initial step in cleanup. While these are vitally important activities, the result is that approximately $110 million was used for "on-the-ground" cleanup at Paducah.

This situation has been exacerbated because we have seen reductions from the Department's requested level of funding. Beginning in fiscal year 1996, Congress—facing its own budget constraints—began appropriating less for the Uranium Enrichment D&D Fund than was requested. Our appropriated levels were less than the levels requested by $10 million, $30 million, $18 million, and $52 million from fiscal year 1996 to fiscal year 1999, respectively, and the funds available for cleanup at Paducah were reduced accordingly. These reductions, coupled with the need to spend funds to maintain the site in a safe condition, have slowed cleanup activities at the site and required us to adjust our priorities and schedules. Working closely with the State and EPA and other stakeholders, we believe we have established and followed a credible strategy and priorities for use of these funds that ensures we are spending our limited funds to the best advantage.

ACTIONS IN RESPONSE TO RECENT INVESTIGATIONS AT PADUCAH

Let me turn now to the some of the specific concerns identified in the Phase I investigation conducted in August 1999 by the Office of Environment, Safety and Health team, and describe the actions we are taking to address those concerns. The Phase I investigation focused on issues from the past ten years and the adequacy of protection provided to workers, the public and the environment today. In addition to examining radiological protection programs, the team also examined environmental conditions and the environmental protection program.

The report on the findings of the Phase I investigation was just released last week. Overall, I have found the report and its conclusions to be fair and accurate. I want to assure you that the Department takes the concerns identified by the investigating team very seriously. We need to correct these conditions. The Manager of the Oak Ridge Operations Office is required to prepare a detailed corrective action plan to address the findings of the report within thirty days of the issuance of the report.

Although we are still evaluating what specific corrective actions are required, I would like to describe the actions we have already taken at the site in response to preliminary findings reported by the review team after the on- site investigation and highlight some additional actions we expect to take.

—In response to early feedback from the investigation team, the Secretary ordered a one-day safety stand-down on September 9, 1999, to emphasize conduct of operations and obtain worker feedback on safety concerns, as well as review the adequacy of radioactive contamination sign postings and other safety measures.

—We have already made changes to improve the sign postings for radioactively-contaminated areas on DOE property. We have, for example, posted signs on both sides of the North-South Diversion Ditch, and at several outfall ditches and culverts associated with Little Bayou Creek.

—The review team found there was a lack of rigor, formality and discipline in the Bechtel Jacobs Company radiation protection program. As I discussed earlier in this statement, we have begun a top-to-bottom review of the radiation control programs at the three gaseous diffusion plants to ensure the controls and procedures are in compliance with DOE requirements and are being fully implemented. The review at Paducah began on October 18 and will be completed at all sites in mid-November, with a report due by the end of November.

—We are also reviewing and making improvements to worker training programs, for example, expanding the information that specifically discusses transuranic materials to the worker radiological worker training.

—The review team found the Department did not have effective oversight of the contractor and its sub-contractors. To address this concern, the Department has assigned two Federal facility representatives to Paducah who will be responsible for monitoring the safety performance of the facility and its operations and will be the primary point of contact with the contractor. Two temporary facility representatives are already in place until permanent full-time employees can be hired. The first permanent position has been filled, and the new facility representative will start on November 7.

—The review team concluded that there has been only limited progress in remediating contamination sources. With the $6 million in additional funding provided for fiscal year 2000, we plan to accelerate the removal of Drum Mountain, pending approval of the necessary documentation by State and EPA regulators, completing removal a year ahead of the previous schedule.

—The review team raised concerns about the shut-down buildings, including whether there were releases of contaminants to the air from the buildings because of deterioration. In response, the roofs of several shutdown buildings will be tested using swipe samples and direct measurements. We will also be conducting a general evaluation of the buildings to determine whether animal infestation or any other pathway is allowing contamination from the buildings to escape that may present a risk to workers or the public and, if necessary, will implement controls.

—The review team concluded that groundwater has not been adequately characterized in some areas. Under the current schedule, we will complete groundwater investigations by August 2000 that will characterize the Southwest plume. However, while not adequate to clearly delineate the leading edge of plumes, we believe the characterization of the Northeast and Northwest groundwater plumes, already approved by the State and EPA, is currently sufficient to determine risks and evaluate cleanup alternatives.

—The review team found that information provided to the public has sometimes been delayed and is not always in forms clearly understood by the public, leading to the perception that the Department is withholding information. While we have already worked to improve communications, there are still opportunities to improve the timeliness and quality of information provided to the public. The contractor and the Oak Ridge Office are jointly preparing a plan for improving communications with the public, which will be issued by November 9. Another DOE field office is also reviewing the public communications documents and process for Paducah, which will provide input to the communications plan.

CONCLUSION

We are making progress at Paducah. Could we make more progress more quickly with more money? Certainly. The same can be said at any of our sites. But, while additional resources would certainly help, the challenges are not solely monetary. Like all of our sites, the problems at Paducah are complex, significant in scale, and technically difficult, and will take time to correct. We will also be evaluating what funding is needed to complete the corrective actions and accelerate cleanup activities to address concerns raised by current and former workers and by the investigation team.

In any event, I will not allow the safety of our workers, the public, or the environment to be knowingly compromised. My first priority for EM is safety—safety of the contractor and Federal workers, and of the people in the communities around our sites. I will hold my managers accountable for ensuring that workers and the public are protected.

STATEMENT OF JIM BICKFORD, SECRETARY OF NATURAL RESOURCES AND ENVIRONMENTAL PROTECTION, COMMONWEALTH OF KENTUCKY

Senator MCCONNELL. Ms. Huntoon, you are out of time.

Secretary Bickford.

Mr. BICKFORD. Thank you, Mr. Chairman. I appreciate the invitation to appear before you today to discuss the issue which you have indicated you are concerned about, as has the Governor and the people of the Commonwealth of Kentucky.

We in Kentucky are very concerned that since the early 1950's the Paducah gaseous diffusion plant has been disposing of and storing radioactive and hazardous waste on site and apparently with very little concern of the eventual environmental consequences. Because the Department of Energy was self-regulating, the Commonwealth was not aware of the extent of the problems until 1996.

CHARACTERIZATION OF SITE CONTAMINATION

Since this time, we have had extreme difficulty getting DOE to define or characterize the extent of the contamination and to take timely action to clean it up. In 1991 and 1992, the Commonwealth and EPA issued permits requiring DOE to clean up over 200 sites at the Paducah facility which contained radioactive and hazardous mixed wastes. DOE resisted these efforts through litigation. We were able to resolve this litigation through an agreed judgment implementing the permits.

However, in 1996 the cabinet issued a permit for a solid waste landfill which restricted the level of radionuclides in solid wastes. DOE resisted this effort and has taken us to State and Federal court, arguing that we do not have the authority to place conditions on solid waste containing radionuclides. The reason stated by DOE is that they are self-regulating under the Atomic Energy Act.

In 1994 the Paducah facility was placed on the EPA Superfund national priority list, as you had mentioned. This list contains the most severely contaminated sites in the United States. As the result of this listing, the cabinet, EPA, and DOE entered into a Federal facilities agreement in 1998. As the Governor mentioned, this agreement requires cleanup by 2010.

To date, very little progress has been made. DOE states that adequate funding has not been made available. We agree with that. From our best estimates, DOE will be in violation of the agreement in a year or so if additional funds are not made available and DOE does not make progress in actually removing the sources of the contamination.

It is extremely difficult for us to estimate total cleanup costs because we do not know the amount and nature of what DOE has buried and disposed of on site since the early fifties. We know that several landfills designated for non-hazardous solid wastes have had hazardous and radioactive wastes placed in them. We know that the area known as Drum Mountain has radioactive waste stored above ground. DOE lists the ground under the site as a burial site, but it is unknown what is buried there. We have also been told that several areas contain classified waste, but to date we do not know what is there.

SITE CONTAMINATION

The point is that there are many areas on site that must be cleaned up. We have no idea what is there and how much it will take to clean it up. Based on what we do know, we have estimated, as the Governor mentioned, that up to $2 billion will be required. That is about $200 million a year if we are going to get it cleaned up in 10 years. Current DOE funding is less than one-fifth of that amount.

We prioritized what we believe must be done at once, that is in the next year. First, begin remediation and removal of radioactive waste burial grounds. There are three plumes of groundwater that are moving off site and have contaminated trichloroethylene and radionuclides that must be contained and treated.

Several drainage ditches and creeks on and off site must be cleaned up. Radioactive tar materials near the landfills must be removed, the black ooze. Non-operational buildings, C–340, C–410, C–420, must be investigated, stabilized, and cleaned up. On and offsite dump sites must be investigated and characterized. Drum Mountain must be removed, material buried under it must be characterized, and the recycling, if necessary or if approved, of scrap metal must be accelerated.

Our best estimate is these activities will cost $646 million over the next 3 fiscal years. That is an average of about $215 million per year and current DOE funding for the cleanup at Paducah has averaged about $39 million for the past decade.

In summary, there is no doubt that a serious cleanup program is required at Paducah. We need adequate funding for the Paducah facility and for DOE to get about cleaning up the site.

STATEMENT OF RICHARD D. GREEN, DIRECTOR, REGION IV, WASTE MANAGEMENT DIVISION, ENVIRONMENTAL PROTECTION AGENCY

ACCOMPANIED BY JOHN JOHNSON, CHIEF, FEDERAL FACILITIES BRANCH, ENVIRONMENTAL PROTECTION AGENCY

Senator McCONNELL. Thank you, Secretary Bickford.

Mr. Green.

Mr. GREEN. Thank you, Mr. Chairman. With me today is John Johnson, Chief of our Federal Facilities Branch.

EPA's role in conjunction with Kentucky at Paducah is to oversee DOE's cleanup. I want to acknowledge the actions taken so far by DOE in response to EPA's 1988 consent order, Kentucky's permit, and the 1998 Federal facilities agreement, FFA. We have worked together to take action with the most pressing areas, the significance of which should not be lost in my remarks today about the actions that are still needed.

I also want to mention that the FFA combines hazardous wastes regulatory——

Senator McCONNELL. You might want to put that mike in front of yourself.

Mr. GREEN. Yes, sir.

I also want to mention that the FFA combines hazardous waste regulatory requirements and Superfund requirements, cuts documentation virtually in half, and provides a regulatory vehicle to accelerate cleanup at the site.

CLEANUP NEEDS

We see the cleanup needs falling in five major areas: one, expansion of the ongoing cleanup of contaminated groundwater; two, cleanup of surface water leaving the site; three, removal or treatment of onsite waste materials; four, decontamination and demolition of deteriorating buildings and other structures; and five, investigation of offsite disposal.

I want to briefly summarize each area. First, groundwater. EPA's evaluation of DOE's data pursuant to the 1988 order has shown that the groundwater, which was at one time a source of drinking water for the nearby residents, is now contaminated. Because of this, this drinking water source is not available for use by the community now or in the foreseeable future.

Currently, DOE is required by the consent order and the FFA to provide the residents in the affected area with clean drinking water. The long-term goal is to return the groundwater source to a usable state as a drinking water source. Continuing and expanding the recovery and treatment of the contaminated groundwater is necessary to meet this long-term goal. As at other Superfund sites, the contamination sources must be eliminated and the groundwater itself must be treated.

Second, surface water. Surface water has been impacted by radioactive contaminants and hazardous substances discharged from the facility and by interaction with contaminated groundwater also. Two creeks flow through the site on their way to the Ohio River. Between the site and the Ohio, they pass through the West Kentucky Wildlife Management Area. Years of plant effluent and other releases from past operations such as spills have caused contamination of sediments and creek banks along these streams and there is evidence that some risk—there is evidence that contaminated groundwater seeps back into the creeks.

People using these waters for recreation are at some risk, according to DOE's latest report, as is the ecosystem itself. The areas of high contamination which are sources for continued releases, such as the North-South Diversion Ditch and Outfall 11 Ditch, should be excavated in order to reduce further contaminant migration and exposure. Additionally, these creeks must be thoroughly surveyed in keeping with DOE's work plan for this, which was submitted last month, and any areas of high contamination must be found and excavated.

Third, onsite waste material. From DOE's latest report, there are about 37,000 cylinders of uranium hexafluoride, or UF6, a highly toxic radioactive substance, stored at outside storage areas around the facility. There is direct gamma radiation coming from each of them. If the cylinders are to be reused, the materials in the cylinders are to be reused, it needs to be done promptly.

Additionally, according to DOE's latest report, there are more than 6,000 cubic meters of low level waste, equivalent to about 31,000 55-gallon drums, stored on site. About 25 percent of this waste is stored in some 8,000 containers outside on bare ground and not covered.

There are numerous burial grounds and huge piles of contaminated scrap metal, such as Drum Mountain. These sources of waste materials continue to leak into the ground and surface waters and should be contained and removed or treated permanently.

Fourth, building decontamination and demolition. There are many unused buildings, storage areas, and other structures on the site, some of which are causing releases of radioactive and hazardous substances. EPA believes that these structures should be decontaminated and demolished as soon as possible.

Fifth and last, offsite disposal investigations. EPA expects DOE will aggressively search out and screen all disposal areas, as well as investigate citizen reports of offsite disposal. If such areas are found, contaminants will need to be characterized and remedies promptly implemented.

In conclusion, the cleanup needs of the Paducah site are extensive, they are very important, and we believe they are for the most part within the range of current treatment technologies.

Thank you, Mr. Chairman.

Senator MCCONNELL. Thank you, Mr. Green.

NEW FUNDING STRUCTURE

Dr. Huntoon, in my opening remarks I stated I thought it would be a good idea if DOE took regulatory and cleanup respects for Paducah and Portsmouth out of the Oak Ridge operations office in order to focus more attention on cleanup. What do you think of that?

Dr. HUNTOON. Well, Senator, I am not for sure that would accomplish the job. We have a new manager at Oak Ridge, Leah Deaver, and she is engaged with us in trying to resolve some of these issues at both Paducah and Portsmouth. I would like to give her the opportunity to try that and see if she can deliver a better management of those two sites.

The reason I hesitate a little bit about starting up more site operations at different places is because that will require more money being put into what we call overhead. We will have to have a larger staff——

Senator MCCONNELL. I think we are all concerned about that. As we all know, you spent $400 million over the last 10 years at Paducah, yet very little actual cleanup has occurred.

I gather you are in the process of developing a new cleanup plan that prioritizes cleanup according to risk. Would that be accurate to say?

Dr. HUNTOON. Yes, sir.

Senator MCCONNELL. Will this new cleanup plan be included in the President's budget for next year?

Dr. HUNTOON. Yes, it will be, but it is also a continuation. I just want to make sure. When we talk about the $400 million that has been spent, a lot of that was the characterization of what is there, the characterization that was needed to work with both EPA and the Commonwealth of Kentucky in setting up these compliances. When someone says compliance they think of like a contract, but actually compliance is getting the work done to meet these agreements.

So we have been doing the characterization and there has been some cleanup accomplished.

Senator MCCONNELL. Yes, but I want to look forward. Will the new cleanup plan be included in the President's budget? Can we expect the President's budget to increase funding for Paducah and Portsmouth? And if so, how much is going to be requested?

Dr. HUNTOON. I do not know that exact number. I know that it does include an increase in this, but it is not as much money as we could use.

Senator MCCONNELL. Well, it is your request. Why do you not ask for as much as you think you need and see what happens?

Dr. HUNTOON. Well, we will do that. We have asked—each year we have asked for the past 4 years, and we have been cut back on our money in the D&D funding.

Senator MCCONNELL. Looking at all that has happened in the last couple of months, why do you not ask for what you need and see what happens.

Dr. HUNTOON. Okay.

DEPLETED URANIUM CYLINDERS

Senator MCCONNELL. Mr. Magwood, could you provide this committee an update on the Department's progress to build a conversion facility to deal with the 60,000 cylinders of depleted uranium?

Mr. MAGWOOD. Yes, Senator, I would be happy to. We do expect to issue a new draft request for proposals to the private sector next month. We will then receive comments from the private sector and issue the final RFP as soon as possible. We are trying very hard to get that done before the end of the year.

One of the issues that is slowing us down a little bit is that, because of these issues that have come up related to the recycled material, it has been necessary for us to go back and search through all the records that exist to confirm that our depleted uranium cylinders do not contain hazardous levels of measurable levels of any transuranic materials.

Senator McCONNELL. Is that the reason the schedule slipped?

Mr. MAGWOOD. That is certainly part of the reason.

Senator McCONNELL. Can you confirm reports that the Administration is considering building only one conversion facility?

Mr. MAGWOOD. No, I can confirm that that is not true at this point.

PADUCAH CLEANUP COST

Senator McCONNELL. That is not the case.

Dr. Huntoon, in my opening statement I referred to a 1996 memo that discussed the Department's desire to fund projects at Oak Ridge that do not pose a risk to worker safety or the environment while DOE neglects Drum Mountain. I was pleased—I think you made some news here a few moments ago. We are going to get Drum Mountain cleaned up in calendar year 2000; did I hear you say that?

Dr. HUNTOON. That is correct, with the additional funds that were made available.

Senator McCONNELL. Good. We will look forward to seeing that done on time.

Governor Patton testified earlier that DOE will need to spend approximately $1.9 billion to address the pending cleanup needs at Paducah. Do your calculations square with his?

Dr. HUNTOON. Senator, the calculations that I have seen to date—and I will go back and look at those again—run the cost up to about $1.2 billion total for Paducah, which is not quite the $2 billion that has been mentioned here. So I do not know the difference in those $1.2 billion versus $2. billion, but I will go examine that.

Senator McCONNELL. According to the recently released phase one study, DOE is in full compliance with the Federal Facilities Agreement negotiated between DOE, EPA, and the State, which establishes DOE's cleanup objectives. The phase one investigation points out that very little cleanup has occurred. Since the Federal Facilities Agreement is not worth the paper it is written on, will you commit to renegotiate a new agreement that actually makes cleanup and not testing its mission?

Dr. HUNTOON. We are meeting with these regulators on November 8 and 9, and I hope we will have a better strategy and some agreements coming out of that meeting.

Senator McCONNELL. You heard David Fuller testify that workers disposed of radioactive waste material around the DOE reservation, much of which remains unidentified. How do you explain that and what are you doing to address it?

Dr. HUNTOON. Well, Senator, I do not know what years he was talking about. I cannot explain it. I think it is regrettable, to say the least.

Senator MCCONNELL. In other words, your answer is it was not on your watch, right?

Dr. HUNTOON. Well, it was years ago. I do not know exactly when the disposal was occurring. We are out there trying now to find out what is in some of these disposal units, and that is one of the things that has gone slowly because the people that are actually doing the work to dig up these sites and try to determine what is there and how to deal with them, we have to protect them also, as well as the environment, from digging up sites.

I do not know how to explain that they were buried. That was a common practice in the past.

ADDITIONAL CLEANUP SITES

Senator MCCONNELL. Well, looking forward, then, looking forward, then, in reviewing the cleanup plans published by the Department I have not found mention of a cleanup strategy for the S and T landfills. Knowing that there is radioactive material in and around these sites, how does this change the overall cleanup strategy?

Dr. HUNTOON. Well, we do have a strategy to deal with those landfills. Originally I think a part of it was to leave them and cap them in place. We are now re-examining that about digging them up, characterizing them, and moving the radioactive containing materials away from the site. That is part of our plan.

Senator MCCONNELL. What about Little Bayou Creek and Big Bayou Creek? What action will be taken to mitigate that contamination?

Dr. HUNTOON. Those areas are being looked at. Of course, the issue of all of our groundwater and surface and subsurface migration into various areas is a concern. We are working on that. We are working it back at the source where this stuff is coming from, trying to deal with that, and we will be trying to clean up these creek beds and all as best we can.

We have got two strategies under way with some new technology on trying to get to the groundwater issues with these plumes that were discussed and a new technology is being applied for the first time down at Paducah and it is doing very well. So we have hopes of dealing with the groundwater and the surface water, particularly the runoffs into these creeks.

BECHTEL-JACOBS CLEANUP CONTRACT

Senator MCCONNELL. The phase one report found a number of problems with the current cleanup contract with Bechtel-Jacobs. My understanding is that this new contracting scheme requires the company to subcontract all the work, which has resulted in several safety lapses. Is the Department considering changing this new contracting scheme to improve worker safety?

Dr. HUNTOON. We have asked Bechtel to come back to us with a plan for improving the safety of the workers and the protection of the workers and the environment, as well as the people around the site. That report is due to us before the end of November.

I am concerned as I read the report, as you are, about the sub-contracting issue, the safety not being transferred, if you will, into all the subcontractors. We have to have a plan brought forward for that.

LAYOFFS AT GASEOUS DIFFUSION PLANTS

Senator MCCONNELL. Considering the fact there are likely to be significant layoffs at the two gaseous diffusion plants next summer, what specific steps are being taken to minimize the job loss and transition as many workers as possible into the cleanup?

Dr. HUNTOON. Well, I know that both the USEC and Bechtel management have been talking about this. I talked with the union workers when I was down at Paducah about their concerns about this area. I asked David Fuller this morning when I saw him had that been progressing in their discussions and he said somewhat, but he was not, as he testified, satisfied with that progression.

So, we will continue working that issue. Part of it has to do, as you know with the movement from the M and I contractor to the subs as they are hiring these people, and these same workers that could be hired for those jobs are working for USEC and may get laid off in the summer. So it is a transition issue that has to be worked between the contractors and the union, I believe.

USE OF PADUCAH CLEANUP FUNDING

Senator MCCONNELL. During Senator Bunning's field hearing in Paducah, Jimmy Hodges, the former DOE site manager, testified that 89 percent of the cleanup dollars spent at Paducah were spent to remain in compliance and not directed toward now cleanup. Recently, you visited Paducah and stated that cleanup "is in the eye of the beholder."

I do not believe your statement rings true, considering the fact that DOE has failed to even identify many of the wastes or go after the most pressing cleanups that continue to contribute to the groundwater pollution. When are we going to get a better return on our investment in Paducah?

Dr. HUNTOON. I am hoping that you will be seeing that in the very near future. As I said in you statement and again to you when we were discussing this a few minutes ago, we had to characterize what was there, because if you go in and just start trying to move things out you can endanger the workers who are actually doing the actual work, and we do not want to do that. We want to protect the safety of the workers.

So the characterization of these various most risky spots had to be our first priority, and we have been doing that. We have done some space storage of some legacy wastes. We have gotten that. We have done the surveillance. We shut down some facilities. We have been operating this landfill trying to get things contained within it. And we have been doing this groundwater work, which has not been a small task as far as trying to understand these plumes as they are moving across the site.

Actually, right now we have 165 wells that we monitor often to make sure that we are trying to contain these plumes. We have two processes of pump and treat that are under way with these ground plumes, and we are also trying a process called "Lasagna" to try to get at the subsurface contamination to stop the source.

So a lot of work is being done in those areas, Senator, that is not terribly visible, but very important as far as we are concerned with the risk to the environment and to people.

Senator MCCONNELL. You know, as you sort of lean back in your chair and think about the last 2 months and the public's reaction beginning with the Washington Post story, it seems to me there has been a tendency on the part of everybody to sort of point the finger at somebody else. Either it is not on my watch or I do not have enough money or it is somebody else's problem.

Let me suggest that it is all of our problem and, regardless of what may have happened in the past, it seems to me the best way to proceed is to quit the finger-pointing, to put in the President's budget request early next year what you think you can usefully spend on both worker safety, monitoring, health concerns, and cleanup, and then we will do our dead level best to get the money for you. I think that people in Paducah are tired of the finger-pointing. They really do not care at this point who is responsible for what.

What they want to know is what kind of condition are we in now if we used to work there or currently work there, what is my physical condition, A; and B, what are we doing to clean this mess up, and what is the soonest we could do this. And it seems to me it is your responsibility to come up with a game plan that gets us there.

Then, if we cannot produce the funds up here, then you have every right to point fingers. Do you not think that is a good way for us to proceed?

Dr. HUNTOON. Yes, sir, it is.

Senator MCCONNELL. Senator Craig has come in and I do not know whether he is here to ask questions or because he is just mesmerized by the subject or what.

Welcome, Senator Craig.

Senator CRAIG. Thank you very much, Mr. Chairman.

I am not mesmerized by the subject. I, like many of us, am focused well on it because of the issue of our Cold War legacy and the responsibility at these laboratories for these cleanups to go forward and to lessen, where it exists, the threat to the public.

We have expended a great deal of money over the last decade and will spend a great deal more, and I think your hearing this morning demonstrates that. So I was really here to listen and to become more focused on Paducah. I have a laboratory in my State, as do other Senators, and we are extremely concerned that they are well managed and that when it comes to waste we handle it.

We have struggled mightily over the years trying to convince the public that we can handle waste appropriately and, as you know, that has been a difficult debate here. Some would prefer that you folks down at Paducah leave it where it is; it is your problem, it is not the country's problem. That is not the case at all. It is a national problem.

I guess my only comment to you, Carolyn, would be, when I am sitting here listening I am trying to make comparisons in my mind of the problem at Paducah relative to other facilities. Am I right to assume that, depending on that facility's role and what it did—how does the Paducah problem compare?

Dr. HUNTOON. Well, Senator Craig, as I mentioned in my statement, I traveled around to most all of our sites that the Office of Environmental Management is trying to remediate, clean up, and contain. I have just been really struck with the amount of work that we have to do out there. Idaho alone has a tremendous groundwater problem, sitting on the aquifer that it is. We have a lot of contaminants in the ground and we are trying, and they are doing a pretty good job of remediating that.

Senator CRAIG. At least we think we have got that one contained until we exhume, of course.

Dr. HUNTOON. Until we exhume it.

Senator CRAIG. Here it does not sound that that is the case.

Dr. HUNTOON. That is the case. There is an estimate of 1.7 trillion gallons of contaminated groundwater from our nuclear war waste in this country, 1.7 trillion gallons—this is across the complex—that we are dealing with at all the sites.

So Paducah has a groundwater problem, but so does Idaho, so does Washington, and Savannah River, and Oak Ridge has a tremendous problem. So we are trying to balance those problems with our resources.

The scrap metal that we see around, we have got 202,000 tons of scrap metal at various sites around the country, and we have got 65,000 tons down at Paducah. Low level waste is all over. We have 8.1 million cubic meters of low level waste around the complex. We have 110,000 cubic meters down at Paducah.

Paducah is serious. Paducah has not been dealt with as it should have been and as we will hope to do in our immediate future. But we have got these same issues at every complex in this country.

Senator CRAIG. Well, I agree with my colleague: Come tell us what you need.

Dr. HUNTOON. Okay.

Senator CRAIG. Let us determine as best we can how to prioritize, of course with your input. But where we have got people at risk, obviously priorities are key.

Dr. HUNTOON. Well, and people ask how do you make these decisions, and we make them based on risk. We try to do that. We try to protect the people, the workers, the people that are surrounding these sites, and the environment, because we have got—just as we are concerned about the plumes at Paducah heading toward the river, we are concerned about what is going on up in Washington heading toward the Columbia River.

Senator CRAIG. That is correct.

Dr. HUNTOON. We are concerned about the aquifer in Idaho, we are concerned about the Savannah River down in South Carolina. So we do try to protect the environment, but we are mostly concerned about the safety of the workers.

Senator CRAIG. Thank you.

Thank you very much, Mitch, Senator McConnell.

Senator McCONNELL. I think what I am going to do here is, I had questions for Secretary Bickford and Mr. Green, which I think I am going to submit to you folks in writing, and if you could get those back in within a couple of weeks I would appreciate it.

It has been a long hearing, but I think very productive, and I want to thank you, Dr. Huntoon, for your candor and underscore what Senator Craig just said. I might say about my friend and colleague Senator Craig, he knows about as much about these issues as anybody in the Senate, maybe more than anybody in the Senate. We tend to look to him when this subject comes up and, even though Paducah is not in Idaho, I wanted to express my gratitude to him for coming by.

It is a huge issue in my State, as you can imagine, and we look forward to getting a request in the President's budget that will give us a chance to make some real progress.

CONCLUSION OF HEARING

I also want to thank you for your commitment to get rid of the drums next calendar year. That would be a visible sign of progress that I think everyone could applaud.

With that, this hearing is concluded. Thank you.

[Whereupon, at 12:13 p.m., Tuesday, October 26, the hearing was concluded, and the subcommittee was recessed, to reconvene subject to the call of the Chair.]

○

APEX
TP_CF
9780160603662

76523826—24